75th Conference
on Glass Problems

75th Conference on Glass Problems

A Collection of Papers Presented at the
75th Conference on Glass Problems
Greater Columbus Convention Center
Columbus, Ohio
November 3–6, 2014

Edited by
S. K. Sundaram

The
American
Ceramic
Society

WILEY

Published by John Wiley & Sons, Inc., Hoboken, New Jersey.
Published simultaneously in Canada.

No part of this publication may be reproduced, stored in a retrieval system, or transmitted in any form or by any means, electronic, mechanical, photocopying, recording, scanning, or otherwise, except as permitted under Section 107 or 108 of the 1976 United States Copyright Act, without either the prior written permission of the Publisher, or authorization through payment of the appropriate per-copy fee to the Copyright Clearance Center, Inc., 222 Rosewood Drive, Danvers, MA 01923, (978) 750-8400, fax (978) 750-4470, or on the web at www.copyright.com. Requests to the Publisher for permission should be addressed to the Permissions Department, John Wiley & Sons, Inc., 111 River Street, Hoboken, NJ 07030, (201) 748-6011, fax (201) 748-6008, or online at http://www.wiley.com/go/permission.

For general information on our other products and services or for technical support, please contact our Customer Care Department within the United States at (800) 762-2974, outside the United States at (317) 572-3993 or fax (317) 572-4002.

Wiley also publishes its books in a variety of electronic formats. Some content that appears in print may not be available in electronic formats. For more information about Wiley products, visit our web site at www.wiley.com.

Library of Congress Cataloging-in-Publication Data is available.

ISBN: 978-1-119-11747-6
ISBN: 978-1-119-11748-3 (special edition)
ISSN: 0196-6219

Printed in the United States of America.

10 9 8 7 6 5 4 3 2 1

Contents

REFRACTORIES

SENSORS AND CONTROL

MODELING

Foreword

The 75th Glass Problem Conference (GPC) is organized by the Kazuo Inamori School of Engineering, The New York State College of Ceramics, Alfred University, Alfred, NY 14802, and The Glass Manufacturing Industry Council, Westerville, OH 43082. The Program Director was S. K. Sundaram, Inamori Professor of Materials Science and Engineering, Kazuo Inamori School of Engineering, The New York State College of Ceramics, Alfred University, Alfred, NY 14802. The Conference Director was Robert Weisenburger Lipetz, Executive Director, Glass Manufacturing Industry Council, Westerville, OH 43082. The themes and chairs of five half-day sessions were as follows:

Glass Melting
Glenn Neff, Glass Service, Stuart, FL, Martin Goller, Corning Incorporated, Corning, NY, and Jans Schep—Owens-Illinois, Inc., Perrysburg, PA

Forming
James Uhlik, Toledo Engineering Company, Inc., Toledo, OH and Kenneth Bratton, Emhart Glass Research Inc., Windsor, CT

Energy and Environmental
Uyi Iyoha—Praxair Inc., Tonawanda, NY and Warren Curtis, PPG Industries, Pittsburgh, PA

Refractories
Laura Lowe—North American Refractory Company, Pittsburgh, PA and Larry McCloskey—Anchor Acquisition, LLC, Lancaster, OH

Sensors and Control
Jans Schep—Owens-Illinois, Inc., Perrysburg, PA and Elmer Sperry, Libbey Glass, Toledo, OH

Modeling
Bruno Purnode, Owens Corning Composite Solutions, Granville, OH and Andrew Zamurs, Rio Tinto Minerals, Greenwood, CO

Preface

Currently, we mark the 75th year of the Glass Problems Conference (GPC). In continuing the tradition that dates back to 1934, this volume is a collection of papers presented at this historic meeting published as the 2014 edition of the collected papers. The manuscripts included in this volume are reproduced as furnished by the presenting authors, but were reviewed prior to the presentation and submission by the respective session chairs. These chairs are also the members of the GPC Advisory Board. I appreciate all the assistance and support by the Board members. The American Ceramic Society and myself did minor editing and formatting of these papers. Neither Alfred University nor GMIC is responsible for the statements and opinions expressed in this volume.

As the Program Director of the GPC, I enjoy continuing this tradition of serving the glass industries. I am thankful to all the presenters at the 75th GPC and the authors of these papers. The GPC continues to grow stronger with the support of the teamwork and audience. I appreciate all the support from the members of the Advisory Board. Their volunteering sprit, generosity, professionalism, and commitment were critical to the high quality technical program at this Conference. I also appreciate continuing support and leadership from the Conference Director, Mr. Robert Weisenburger Lipetz, Executive Director of GMIC. I look forward to continuing our work with the entire team in the future.

S. K. SUNDARAM
Alfred, NY
December 2014

Acknowledgments

It is a great pleasure to acknowledge the dedicated service, advice, and team spirit of the members of the Glass Problems Conference Advisory Board in planning this Conference, inviting key speakers, reviewing technical presentations, chairing technical sessions, and reviewing manuscripts for this publication:

Kenneth Bratton—*Emhart Glass Research Inc. Hartford, CT*
Warren Curtis—*PPG Industries, Inc., Pittsburgh, PA*
Martin Goller—*Corning Incorporated, Corning, NY*
Uyi Iyoha—*Praxair Inc., Tonawanda, NY*
Robert Lipetz—*Glass Manufacturing Industry Council, Westerville, OH*
Laura Lowe—*North American Refractory Company, Pittsburgh, PA*
Larry McCloskey—*Anchor Acquisition, LLC, Lancaster, OH*
Glenn Neff—*Glass Service USA, Inc., Stuart, FL*
Bruno Purnode—*Owens Corning Composite Solutions, Granville, OH*
Jans Schep—*Owens-Illinois, Inc., Perrysburg, PA*
Elmer Sperry—*Libbey Glass, Toledo, OH*
Phillip Tucker—*Johns Manville, Denver, CO*
James Uhlik—*Toledo Engineering Co., Inc., Toledo, OH*
Andrew Zamurs—*Rio Tinto Minerals, Greenwood, CO*

Glass Melting

EFFECT OF DISSOLVED WATER ON PHYSICAL PROPERTIES OF SODA-LIME-SILICATE GLASSES

Udaya K. Vempati
Research and Development, Owens-Illinois, Inc.
One Michael Owens Way, Perrysburg, OH 43551, USA

Terence J. Clark[1]
Bowling Green, OH 43402, USA

ABSTRACT

Dissolved gases in glass melts are known to influence properties of the melts as well as the resulting glass and dissolved water is thought to be one of the most influential of all the dissolved gases. In this work, the effect of vacuum processing and the ensuing changes in dissolved water concentration on various physical properties of soda-lime-silica glasses were studied. Glass melts with varying dissolved water concentration were prepared by melting frit at atmospheric and sub-atmospheric (≈ 100 torr) pressures at 1450°C. The physical properties of these melts and the resulting glasses were determined by rotating spindle viscometry, beam bending viscometry, and spectroscopy. The densities of the glass samples were also determined. Results from these experiments are discussed in relation to prior work in the field and the implications of changes in properties on the glass making process are discussed.

INTRODUCTION

The presence of dissolved gases and their influence on properties of soda-lime-silica, as well as other silicate melts and glasses, is well known and characterized in detail [1–10]. A common route by which gases are introduced in glasses is through batch constituents, which may contain both physically adsorbed and chemically bonded gases. For example, both limestone ($CaCO_3$) and soda ash (Na_2CO_3), commonly used in the synthesis of soda-lime-silica glasses, release CO_2 during melting of the glass batch. Much of this gas escapes the melt in the form of bubbles during melting and fining, but some can remain dissolved in the melt. Depending on the raw materials used, many other gases can be present in a soda-lime-silica melt including H_2O, SO_2, O_2, NO, and NO_2. These gases may also remain in the resulting solidified glass. When the glass melt is subject to sub-atmospheric pressure or vacuum, the dissolved gas concentration is expected to change. Consequently, the physical and chemical properties of glasses obtained from vacuum-processed melts can be expected to be different from corresponding glasses obtained from atmospheric or ambient melts. Variations in atmosphere surrounding the melt, particularly in the oxygen content, could also influence the properties of glasses from vacuum-processed melts.

Several earlier reports have discussed the influence of vacuum processing on the properties of silicate glasses. For instance, Fenstermacher et al. [2] measured the dissolved water concentration and viscosity of soda-lime silicate glasses obtained from glass batches melted at both ambient pressure and under vacuum. They found that the softening point of glass samples resulting from vacuum melting can be as much as 5°C higher than the glasses obtained from ambient melts and attributed this difference to the lower dissolved water content in the vacuum melt samples. Similarly, Graff and Badger [9] measured the viscosity of glasses (presumably soda-lime silicates) obtained from melts subject to vacuum, as well as those saturated with carbon dioxide and water, and found that, at a given temperature, the samples from vacuum

---

[1] Work performed while employed at Owens-Illinois, Inc.

melts generally had higher viscosities than those from saturated melts. Sproull and Rindone [11] melted lithium-potassium silicates under vacuum and found that fibers drawn from these vacuum melts have lower strength, by as much as 25%, than those produced from ambient melts. They attributed the reduction in strength to a heterogeneous microstructure brought about by low levels of oxygen in the atmosphere surrounding the melt under vacuum.

Most of this early work focused primarily on a specific property of the glass, such as viscosity or mechanical strength. However a more thorough and comprehensive investigation of the effects of vacuum processing on various properties of soda-lime silica glasses and melts is lacking. To address this gap, viscosity, density, Fourier transform infrared (FTIR) and ultraviolet (UV)-Visible (Vis) transmission spectra of soda-lime silicate glasses obtained from vacuum melts were measured in the present study. These data were compared with the corresponding properties of glasses produced by standard practice, i.e. from melts prepared at atmospheric pressure. Any difference in properties between the two sets of glasses was explained in context of differences in dissolved water concentration and/or processing glass under an oxygen-deficient atmosphere. It should be pointed out that although many different gases can be dissolved in a silicate glass, the focus of this work was limited to water, *because water is thought to be the dominant species among dissolved gases in soda-lime-silica glasses* [12].

The central finding of this work was that vacuum processing has a measureable impact on the viscosity and UV-Vis transmission spectra of a soda-lime-silica glass. Specifically, at viscosities corresponding to melting and gob formation, the temperatures of vacuum processed melt can differ by about 10°C from the ambient melt. This result is consistent with those previously reported by others and can be attributed to the lower dissolved water concentration in the vacuum melt [2]. The dominant wavelength in the UV-Vis transmission spectrum of the vacuum processed flint glass is found to shift to lower values compared to a standard flint glass. Changes in redox brought about by the vacuum processing may have some bearing on this result. However, the density of the glass was not effected by vacuum processing. Mechanical properties, critical from an end use viewpoint, have not been characterized here and remain a topic for future work.

EXPERIMENTAL METHODS

Details of sample preparation and data collection methods are provided in this section. Note that x-ray fluorescence (XRF) measurements were carried out to establish composition of a vacuum processed glass sample. Within experimental error, the composition of this sample was found to be similar to a sample prepared by standard methods. Hence, except for their dissolved gas content, both kinds of glasses are treated to be of similar chemical composition in this work.

Batching and melting

Soda-lime-silica glass frit was prepared by melting batch materials in an electric furnace at ambient pressure. Raw materials from bulk suppliers were used in preparing glass batches of nominal composition, in mass fraction, 73.7% SiO_2 - 13.6% Na_2O - 11.3% CaO - 1.4% Al_2O_3. The frit was then melted for 2.5 h in an evacuated furnace, where the pressure was maintained between 13.3 kPa and 17.3 kPa. This is referred to as "vacuum-processed glass". The frit was also melted in ambient, i.e. at 101.3 kPa, for 2.5 h to obtain glass samples that are representative of glasses produced by existing commercial practices. This is the "ambient-processed glass".

FTIR, UV-Vis transmission, and density measurements

Samples for FTIR experiments were prepared as described above by first melting ≈300 g of frit at 1450°C for 2.5 h either in ambient or in vacuum. These melts were then poured onto a room temperature metal plate and pressed into discs. The discs were annealed at 550°C for 10 min and subsequently slow cooled to room temperature. They were then cut to size (3.81 cm diameter) using a core drill and polished on both sides. The polished discs were placed in an FTIR spectrometer sample chamber and purged with nitrogen gas for 15 min before collecting absorption spectra. A background scan was also collected with an empty sample chamber after a fifteen minute purge with nitrogen gas. Absorption spectra were used to calculate the dissolved water concentration in glass samples as will be described in the results section below.

Glass prisms measuring 35 mm × 35 mm × 13 mm for UV-Vis transmission were also prepared from the melts used for producing the FTIR samples. A small section, roughly 2 mm thick, was cut from the 35 mm × 13 mm face of the prism using a diamond saw. The cut faces were ground and polished to a mirror finish. A UV-Vis spectrophotometer was used for the transmission measurements.

Density measurements were carried out on crushed glass pieces obtained from the same sample set as used for FTIR and UV-Vis spectroscopy measurements by the Archimedes method. Deionized water with a few drops of surfactant was used as the weighing medium. A wire mesh, hung from a support frame connected to a balance, was used to suspend the samples in the water. A thermometer was clipped to the water beaker to continuously monitor the water bath temperature. Six glass samples were used from two different batches at each condition, vacuum and ambient processing, for the density measurement.

Viscosity

Viscosity measurements were carried out by two different methods. A rotating spindle viscometer was used to measure viscosities in the (10^1-10^4) Pa·s range, whereas a beam bending viscometer was used for measurements in the (10^8-10^{13}) Pa·s range. Combining data from these two measurement series, viscosity of the glass in the (10^1-10^{13}) Pa·s range was obtained by interpolation.

For the rotating spindle viscometer experiments, approximately 320 g of frit was melted in a platinum crucible for 2.5 h at 1450 °C in the viscometer furnace. The furnace temperature was then set to desired value and the rotating viscometer spindle was lowered into the melt. The torque experienced by the spindle was measured and noted at defined time intervals. The viscosity of the melt was calculated from the measured torque and spindle RPM and plotted as a function of time. A steady state value of this measurement provided the viscosity of the melt at that temperature. This process was repeated for different temperatures. The procedure for the vacuum processed samples was slightly different. Approximately 320 g of frit was melted in a platinum crucible for about 2.5 h at 1450°C in a laboratory furnace evacuated to between 13.3 kPa and 17.3 kPa. After the 2.5 h melting the crucible was transferred to the viscometer furnace. Viscosity measurements were then carried out using the procedure described above for the ambient sample.

In case of beam bending viscometer experiments, samples were prepared from melts conducted by heating about 200 g of frit at 1450 °C for 2.5 h in platinum crucibles. The melting was either in vacuum or at ambient pressure. After 2.5 h soak, the melts were poured into rectangular cavities measuring 90 mm × 50 mm × 10 mm. The plates were machined into beams measuring 55 mm × 4 mm × 4 mm and then loaded, one at a time, into a beam bending viscometer for measuring the viscosity. Measurements were carried out in accordance with ASTM standard C-598. Both sets of viscosity measurements were repeated on two different batches of material, with the average of each measurement reported below.

RESULTS AND DISCUSSION

FTIR spectroscopy

FTIR spectroscopy was used in this work to quantify the concentration of dissolved H_2O. The FTIR absorption spectra of ambient and vacuum processed glass samples are shown in Figure 1. Bands at about 2800 and 3500 cm^{-1}, that are generally attributed to OH groups [3], are clearly visible. Absorption in ambient processed glass samples is found to be generally higher than in the vacuum processed samples due to the higher dissolved water content. The relationship between infrared absorption and dissolved water concentration C_i, is given by the Beer-Lambert equation:

$$C_i = \frac{A_i}{\rho \cdot t \cdot \varepsilon_i}$$

where A_i is the absorption of ith band, ρ is density, t is thickness, and ε_i is the absorption coefficient corresponding to the ith band of the sample. The band corresponding to 2800 cm^{-1} was used to calculate the concentration of the dissolved water. Behrens and Stuke [13] have shown that the practical molar absorption coefficient corresponding to this band is independent of the dissolved water concentration in soda-lime-silica glasses. Hence this band was used instead of the more commonly used 4500 cm^{-1}, 5200 cm^{-1}, or the 3550 cm^{-1} bands. The practical molar absorption coefficient for the 2800 cm^{-1} band was taken from Behrens and Stuke [13] as 50.8 l·mol^{-1}·cm^{-1}.

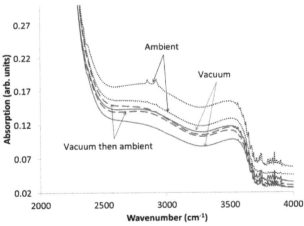

Figure 1. Infrared absorption spectra of ambient and vacuum processed glass samples. Also shown are the absorption spectra from samples that have been vacuum processed and subsequently melted under ambient conditions. Note that the spectra of these samples is similar to those of vacuum processed samples.

Assuming a density of 2.5×10^{-3} kg/m³, the dissolved water concentration (in mass fraction) calculated using Eq. (1) was found to be $0.0237 \pm 0.0027\%$ in the ambient processed glass and $0.0194 \pm 0.0015\%$ in the vacuum processed glass. The density of both ambient and vacuum processed glasses was assumed to be the same for this calculation. This assumption is reasonable as will be shown below. The dissolved water concentration values obtained in this work are in reasonable agreement with those found by Fenstermacher et al. [2] and by Jewell et al. [3].

Infrared absorption spectra were also collected from glass samples obtained by re-melting vacuum processed glass in ambient (for 2.5 h). The absorption spectra of the vacuum processed glass and vacuum processed then ambient melted glass were found to be similar. This result suggests that once water has been extracted from the glass, melting under ambient conditions subsequently has little effect on the dissolved water concentration, at least within the time scale of the experiments. This result provides confidence in asserting that the measurements of physical properties of vacuum processed glasses, which are carried out under ambient conditions, are representative of the true properties of the vacuum processed glass.

Viscosity

Results from the viscosity experiments are shown in Figure 2. The data from the ambient and vacuum processed samples can be approximated by the Fulcher equation $\log \eta = A + B/(T - T_0)$, where η is the viscosity, T is the temperature, and A, B, and T_0 are constants, which are used as fitting parameters. Although it appears that the viscosities of the two kinds of glasses is similar, a close observation reveals systematic differences.

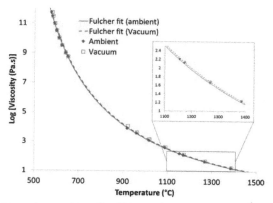

Figure 2. Viscosity versus temperature of ambient- and vacuum-processed glasses. A least squares fit to the data gave the parameters of the Fulcher equation. The resulting curves are shown as lines.

In Table 1, temperatures corresponding to some viscosities of interest in glass manufacturing operations are listed for both ambient-processed and vacuum-processed glass. In general, the temperature corresponding to a particular viscosity is higher for vacuum-processed glass than for a comparable ambient-processed glass. The difference between the temperatures is largest at low viscosities and becomes smaller as the viscosity increases.

The softening point, defined as the temperature corresponding to a viscosity of $10^{6.65}$ Pa·s, is about three degrees higher in vacuum processed glass (735°C) compared to the ambient processed glass (732°C). The gob temperature, corresponding to a viscosity of 10^2 Pa·s, is 1192°C for the ambient processed glass and 1200°C for the vacuum processed glass. An 8°C difference in gob temperatures may necessitate changes in production practices. Hence this result should be carefully considered when evaluating the use of vacuum or sub-atmospheric pressures in glass melting.

The increased viscosity of vacuum processed glass compared to ambient processed glass at a given temperature can be explained by the decrease in the dissolved water content. As the dissolved water content reduces, the glass network becomes more cross-linked and the viscosity increases. If the water is only dissolved chemically, i.e. as OH species, then the isokom temperatures corresponding to lower viscosities are more influenced, as observed in this work and that of Fenstermacher et al. [2]. Stuke et al. have shown that below 0.5 wt%, water is dissolved primarily in the form of OH groups in soda-lime-silica glasses. When the water is dissolved primarily as H_2O molecules, as is the case above 3 wt%, then the low temperature viscosity may be more influenced. This is observed in Del Gaudio et al.'s work [14], in which they developed a parametric model to predict viscosity of soda-lime-silica glasses of different dissolved water concentrations. This model predicts that the viscosity difference between float glasses of different dissolved water concentrations should be greatest at low temperatures. The reason for such prediction may be that their parametric model is developed for glasses with larger dissolved water concentrations, where the water is thought to exist primarily as H_2O molecules and not as OH species. It should also be pointed out that Jewell et al. [3] have found similar trends in viscosity as those reported here and by Fenstermacher et al. [2].

Table 1. Temperatures of the ambient- and vacuum-processed glasses corresponding to a specific viscosity value.

Log[Viscosity (Pa·s)]	Temperature (°C)	
	Ambient-processed	Vacuum-processed
1	1451	1461
2	1192	1200
6.65	732	735
11	588	590
12.4	560	562

UV-Vis transmission

In terms of visual appearance, the ambient and vacuum processed glasses were found to be similar. However, subtle differences were noticed in their transmission spectra. The transmission spectra were collected on rectangular parallelepiped samples over a wide spectrum of wavelengths, ranging from 300 nm to 1100 nm. Results from the transmission experiments are shown in Figure 3, where normalized transmission is plotted against the wavelength of incident light. Solid lines represent data collected from vacuum processed glass and dashed lines from ambient processed glass. The transmission data were normalized by dividing the absolute transmission by the transmission at most intense wavelength. The wavelength corresponding to the maximum transmission (a value of 1 on the curve) is referred to as dominant wavelengths in this work. The dominant wavelength between 520 nm and 590 nm in the vacuum processed samples ranged slightly lower than the ambient processed samples, where the dominant wavelength ranged between 560 nm and 590 nm as shown in the inset in Figure 3.

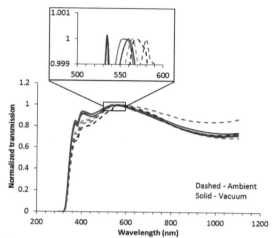

Figure 3. Transmission spectra of ambient and vacuum processed glasses. Absolute transmission was divided by the maximum transmission to obtain the normalized data.

The reason that a shift in the dominant wavelength of vacuum processed glasses is observed may lie in the fact that these glasses were processed under reduced amounts of oxygen compared to the ambient processed glass. This lower oxygen content may lead to higher FeO in vacuum processed glasses and hence a shift of the dominant wavelength to lower values. Some support for this argument is found in the UV-Vis transmission spectra of the glasses shown in Figure 3. Strong absorption below 370 nm is generally attributed to Fe^{3+}, which is found to be more prominent in the ambient processed glasses than the vacuum processed glasses. In other words, the lower amount of Fe^{3+} in the vacuum processed glasses lowers the absorption below 370 nm and shifts the dominant wavelength to lower values. Weyl ([15], see page 105) discusses similar changes in absorption spectra observed by Andresen-Kraft [16] in sodium trisilicate ($Na_2O \cdot 3SiO_2$) glasses of varying iron compositions. Andresen-Kraft found that when a large amount of iron is added to the glass, majority of iron exists in the 3+ or Fe_2O_3 state and the glass has a greenish-brown color. On the other hand, at low iron content, some FeO is present and leads to a light green colored glass, suggesting a shift in the dominant wavelength.

If the shift in dominant wavelength is truly due to the changes in redox, it would then follow that the 2.5 h melting done in ambient/vacuum was long enough to cause oxygen to diffuse through the melts and alter the redox equilibrium. Goldman and Gupta [17] have shown that Fe^{2+}/Fe^{3+} decreases by about a factor of 3 over a three hour period in a calcium aluminum borosilicate glass melt subjected to oxidation. Oxygen diffusion coefficient in this melt was found to be on the order of 10^{-7}, which is similar to that expected in a soda-lime-silica melt [18]. Hence it is plausible that the reverse effect can be expected when a glass is melted in an oxygen deprived environment. The change in Fe^{2+}/Fe^{3+} was not determined in this work and remains a subject for future studies.

Density

The variation in the concentration of dissolved gases due to vacuum processing may be expected to result in changes in density of the glass specimen. While the density of vacuum-processed glass was found to be slightly lower than the ambient-processed glass, as seen in Figure 4, the difference was smaller than the error in measurement. The average density (over twelve different measurements) in both cases was found to be 2.50 ± 0.03 g/cc. Fenstermacher et al. [2], reported a similar result. It should be noted that Shelby [19] found the density of a silica glass to vary as much as 0.15% when the dissolved water concentration changed from 0 to 800 ppm by weight. However, the changes in dissolved water concentration in the glasses studied here were much smaller and the expected change, if any, in density would be smaller than the error in measurement.

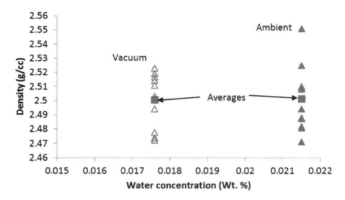

Figure 4. Density of ambient (open symbols) and vacuum (filled symbols) processed glass samples as a function of dissolved water concentration.

CONCLUSION

Vacuum processing and the ensuing changes in dissolved water concentration are found to have some effect on the viscosity and transmission spectra of soda-lime-silicate glasses. Specifically, the temperature corresponding to viscosities in the 10^1–10^3 range can increase by up to 10°C as the dissolved water concentration decreases from 0.0237% to 0.0194%. Such a change (in absolute terms) may necessitate minor changes to glass manufacturing operations and hence should be monitored when vacuum processing or other methods impact water content in the melt are employed. The dominant wavelength in the optical transmission spectra is also found to shift to lower wavelengths in glass samples with a lower dissolved water content. On the other hand, the concentration of dissolved water or the vacuum process has minimal influence on the density of the glass samples.

ACKNOWLEDGEMENT

This work was funded by the R&D organization at Owens-Illinois, Inc. Kathryn Perkins, Faith Workman, and Brett Hixson assisted in sample preparation and with several experiments. Daniel Ragland helped with the FTIR experiments. Carl Fayerweather provided support for the rotating spindle viscometer experiments. Richard Sipes help with UV-Vis experiments is greatly appreciated. Discussions with the O-I's Glass Science Discipline members, both past and present, were helpful in the development of this project and in the preparation of this document. Comments from several members of O-I R&D, particularly Scott Weil, William Pinc, Carol Click, and Ludovic Valette, improved the quality of this document and are greatly appreciated.

Special thanks to Carol Click for presenting this work at the 75th Glass Problems Conference held in Columbus, Ohio.

REFERENCES

[1] Washburn, Footitt and Bunting, "Dissolved gases in glass," *Journal of the Franklin Institute,* vol. 191, no. 6, p. 842, 1921.

[2] J. E. Fenstermacher, R. C. Lesser and R. J. Ryder, "A study of water content of container glasses," *Glass Industry,* pp. 518-521, 1965.

[3] J. M. Jewell, M. M. Spess and J. E. Shelby, "Effect of water concentration on the properties of commercial soda-lime-silica glasses," *Journal of The American Ceramic Society,* vol. 73, no. 1, p. 132–135, 1990.

[4] R. G. C. Beerkens and J. v. d. Schaaf, "Gas release and foam formation during melting and fining of glass," *Journal of the American Ceramic Society,* vol. 89, no. 1, pp. 24-35, 2006.

[5] A. Stuke, H. Behrens, B. C. Schmidt and R. Dupree, "H2O speciation in float glass and soda lime silica glass," *Chemical Geology,* vol. 229, p. 64–77, 2006.

[6] J. E. Shelby and G. McVay, "Influence of water on the viscosity and thermal expansion of sodium trisilicate glasses," *Journal of Non-Crystalline Solids,* vol. 20, pp. 439-449, 1976.

[7] F. W. Kramer, "Solubility of gases in glass melts," *BERICHTE DER BUNSEN-GESELLSCHAFT-PHYSICAL CHEMISTRY CHEMICAL PHYSICS,* vol. 100, no. 9, pp. 1512-1514, 1996.

[8] E. Bourgue and P. Richet, "The effects of dissolved CO2 on the density and viscosity of silicate melts: a preliminary study," *Earth and Planetary Science Letters,* vol. 193, p. 57–68, 2001.

[9] W. A. Graff and A. E. Badger, "Viscosity of glass as affected by dissolved gases," *Journal of the American Ceramic Society,* p. 220, 1946.

[10] M. Tomozawa, "Water in glass," *Journal of Non-Crystalline Solids,* vol. 73, pp. 197-204, 1985.

[11] J. F. Sproull and G. E. Rindone, *Journal of The American Ceramic Society,* vol. 57, no. 4, p. 160–164, 1974.

[12] Schloze, "Nature of water in glass, I," *GLASTECHNISCHE BERICHTE,* vol. 32, pp. 81-88, 1959.

[13] H. Behrens and A. Stuke, "Quantification of H2O contents in silicate glasses using IR spectroscopy: a calibration based on hydrous glasses analyzed by Karl-Fischer titration," *Glass science and technology,* vol. 76, no. 4, pp. 176-189, 2003.

[14] P. Del Gaudio, H. Behrens and J. Deubener, "Viscosity and glass transition temperature of hydrous float glass," *Journal of Non-Crystalline Solids,* vol. 353, p. 223–236, 2007.

[15] W. A. Weyl, Coloured Glasses, Sheffield: The Society of Glass Technology, 1951.

[16] Andresen-Kraft, *Glastechnische Berichte,* vol. 9, pp. 577-597, 1931.

[17] D. Goldman and P. Gupta, "Diffusion-controlled redox kinetics in a glass melt," *Journal of the American Ceramic Society,* vol. 66, pp. 188-190, 1983.

[18] R. Doremus, "Diffusion of oxygen from contracting bubbles in molten glass," *Journal of the American Ceramic Society,* vol. 43, pp. 655-661, 1960.

[19] J. E. Shelby, "Density of vitreous silica," *Journal of Non-Crystalline Solids,* vol. 349, pp. 331-336, 2004.

[20] J. F. Sproull and G. E. Rindone, "Effect of melting history on the mechanical properties of glass: I, role of melting time and atmosphere," *Journal of the American Ceramic Society,* vol. 57, no. 4, pp. 160-164, 1973.

COMPARISON OF SEM/EDX ANALYSIS TO PETROGRAPHIC TECHNIQUES FOR IDENTIFYING THE COMPOSITION OF STONES IN GLASS

Brian Collins, Gary Smay, and Henry Dimmick
Agr International, Inc.
Butler, PA 16001, USA

ABSTRACT
       Stones are a problem that can adversely affect glass production of container, flat glass, specialty, or fiberglass manufacturing. Consequently, it is important to quickly and correctly identify the source of the stone and implement appropriate corrective actions. Historically, the analysis of stones required time-consuming petrographic analyses. Recently, glass technologists have adopted a method of analyzing stones consisting of a scanning electron microscope (SEM) together with the use of energy-dispersive x-ray analysis (EDX). These current methods have the potential to provide accurate, detailed information about the stone in a more expedient fashion than typical petrographic analyses. This paper compares data derived from analyses of stone identification using an SEM/EDX to the results obtained from petrographic techniques.

INTRODUCTION:
       Stones are unmelted or re-crystallized solid materials in the amorphous glass matrix (1-4). Stones in the glass articles range from cosmetic issues to sources for possible breakage and can cause production losses. Outbreaks of stones often occur without warning and can either end very quickly or last for several days.
       Sources of stones typically include devitrification, raw material contaminants, refractory materials from erosion of furnace components, and cullet contaminants. In many cases, stones change crystalline phases when elevated furnace temperatures are encountered. Primary stones have at least a portion of the original inclusions present, and have not substantially changed in terms of phases. Secondary stones have melted and then recrystallized to form a different phase compared to the primary stone (2,3). Since the sources of stones can be anything from a mined contaminant, to a misplaced coffee cup in a recycle container, methods need to be in place to identify the stone in order to determine the source.
       This paper compares the data obtained from thirty stones typical of batch materials, refractory components, contaminants, and devitrification.   The study included a visual assessment of the stones coupled with petrographic analysis and techniques using a scanning electron microscope (SEM) equipped with an energy dispersive x-ray spectrometer (EDX) (5,6).

METHODS OF ANALYSIS AND EXPERIMENTAL PROCEDURE
       A visual assessment was made of the stones to characterize the color, melting condition, surface texture, or other physical features that could aid in their identification.

Petrographic Analysis
       After the visual assessment, the stones were cut in half. One half of each stone was set aside for SEM/EDX analysis. The second half was used to prepare a thin section (approximately 30 microns in thickness (2)) that reveals the crystal morphology for use in the petrographic analysis, as shown by the examples in Figure 1. A polarizing microscope with a 1st order red tint

plate was used to determine the extinction angle, sign of elongation, relative birefringence, color, and refractive index of crystals in the thin sections. Determination of these characteristics is beneficial for the identification of the composition of the stones.

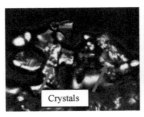  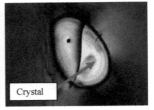

Figure 1. Examples of Crystal Structures Obtained from Petrographic Analyses

SEM/EDX Analysis

A SEM consists of a focused ion beam of electrons directed to interact with the sample specimen, as shown in Figure 2 (5, 6). The "Sample" in the figure is the stone itself, or specific locations within the stone. Typically, the three signals from the interaction of the electron beam and the sample that are of primary interest are the secondary electrons, backscattered electrons, and the characteristic x-rays (5, 6). The emitted electrons create an image of the stone and energy dispersive x-ray analysis is utilized to identify the elemental composition of various components within the stone.

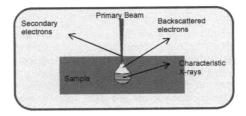

Figure 2. Signal Generation from Surface Analysis by SEM

In the current study, the halves of the stones that were reserved for SEM/EDX analyses were polished to a high gloss with decreasing grit paper and alumina powder. The samples were then mounted for SEM analysis. The SEM/EDX data was in the form of photomicrographs of the surface details and x-ray spectra of areas of interest (e.g., grains, crystals). The SEM employed in this study has an electron beam that can be decreased to a spot size as small as 1.5 microns in diameter. This allows for the identification of the composition (qualitatively) for the majority of crystal sizes and types. The depth of penetration of the electron beam for x-ray analysis is typically 1 - 5 microns, depending on the sample density and the accelerating voltage setting. Thus, the SEM/EDX generates electronic signals from the surface layer of the stone, which provides comparable information as observed in a thin section.

SUMMMARY OF RESULTS
1. SEM/EDX analysis techniques successfully identified all thirty of the stones.
2. EDX analysis definitively determined the composition of all the components of the stones.
3. SEM/EDX is a viable alternative to petrographic analysis and in certain cases increases the amount of information available from the stones.

DISCUSSION OF RESULTS
Thirty stone samples were analyzed in this study. These samples were typical of stones from batch materials, refractory components, raw material / batch contaminants, and devitrification. Eight representative stones were chosen from the thirty samples for detailed discussion and comparison of techniques in this paper.

Case Study #1 - Stone Sample #1
The stone was white in color with a granular structure and a small solution sac (indicating a small amount of melting into the surrounding glass), as shown in Figure 3(a).
Petrographic analysis revealed the presence of grains and laths, as shown in Figure 3b. Based on the physical characteristics, the extinction angle, and the sign of elongation the laths were identified as tridymite. The grains showed a moderate level of birefringence, which in addition to being associated with tridymite, indicated they were likely quartz. SEM EDX revealed the stone consisted of grains and extending laths (3(c)) both composed of silica (3(d)).
The SEM/EDX provided the same information as the petrographic techniques plus it provided the elemental composition of the grains and laths. The stone was primarily quartz grains that were partially converted to a secondary recrystallized lath form (tridymite).
The source of the stone was unmelted silica batch material.

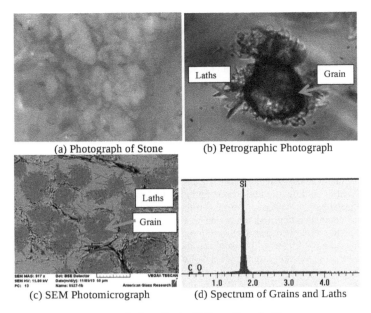

(a) Photograph of Stone (b) Petrographic Photograph

(c) SEM Photomicrograph (d) Spectrum of Grains and Laths

Figure 3. Case Study #1 Stone Sample #1

Case Study #1 - Stone Sample #5

The stone was primarily opaque with evidence of a large solution sac (indicating melting into the surrounding glass) and crystals that appeared to branch, as shown in Figure 4(a). Petrographic analysis revealed crystals with 90° branching and small grains, as shown in Figure 4(b). The physical characteristics, extinction angle, and sign of elongation indicated that the crystals were cristobalite. The grains showed a moderate birefringence, which in addition to being associated with cristobalite, indicated they were likely quartz.

SEM analysis also revealed crystals showing 90° branching and a small number of grains (Figure 4(c)). The EDX analysis revealed the crystals and the grains were composed of silica, as shown in Figure 4(d). The SEM/EDX gave the same information as the petrographic techniques plus it provided the elemental composition of the grains and branching crystals. The stone was quartz that was nearly all converted to cristobalite. The source of the stone was likely silica scum that had recrystallized.

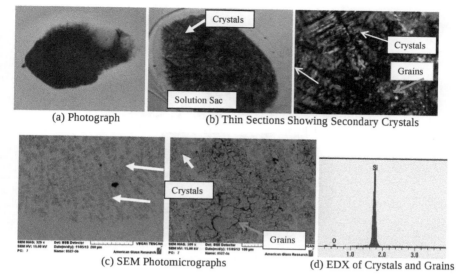

(a) Photograph   (b) Thin Sections Showing Secondary Crystals

(c) SEM Photomicrographs   (d) EDX of Crystals and Grains

Figure 4. Case Study #1 Stone Sample #5

Conclusion from Case Study #1

SEM/EDX analysis provided the same identification of the stones as with the petrographic techniques. Stone sample #1 exhibited primary grains of quartz with a small amount of conversion to tridymite laths. Stone sample #5 was largely a secondary stone composed of cristobalite. Both techniques were able to determine the composition of the stones and data from both techniques led to the same conclusion regarding the source.

Case Study #2 - Stone Sample #11

The stone appeared black, was rounded, with no evidence of a solution sac, as shown in Figure 5(a). The petrographic analysis revealed a solid inner core with a crystalline perimeter (Figure 5(b)). Neither, the inner core of the stone nor the crystal plates at the perimeter could be identified through the analysis methods of crystal morphology. However, based on the physical appearance of the stone when viewed in cross-section, a comparison to the literature identified this type of stone as typical of an iron chromite inner core and perimeter crystals of chromic oxide.

SEM showed a solid inner core with a crystalline perimeter, as shown in Figure 4(c). EDX analysis revealed the inner core was composed of chromium, aluminum, magnesium, iron, and oxygen (Figure 5(d)). The perimeter of the stone (Figure 5(e)) was composed of chromium and oxygen. SEM/EDX analysis of the stone revealed the combination of elements typical of iron chromite. The petrographic techniques were unable to show definitive results as to the composition. However, through the use of reference materials combined with an experienced microscopist the composition of the stone can be identified. The source of the chromite was likely a raw material contaminant.

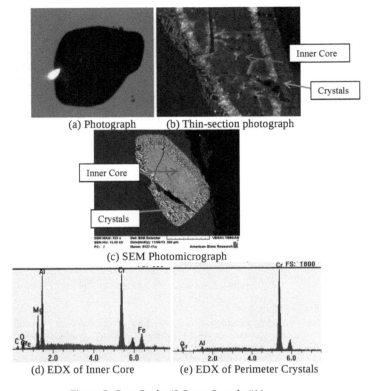

(a) Photograph   (b) Thin-section photograph

(c) SEM Photomicrograph

(d) EDX of Inner Core   (e) EDX of Perimeter Crystals

Figure 5. Case Study #2 Stone Sample #11

Case Study #2 - Stone Sample #24

The stone was opaque and black in color with a tint of green at the perimeter. There was no evidence of a solution sac as shown in Figure 6(a). Petrographic analysis revealed a thin crystalline matrix, as shown in Figure 6(b) that could not be positively identified. However, based on the physical appearance, references in the literature identified this type of stone as chromic oxide crystals. SEM analysis showed that the stone was an agglomeration of thin crystals (Figure 6(c)). EDX revealed (Figure 6(d)) the crystals were composed of chromium and oxygen.

SEM/EDX analysis of the stone revealed the composition as chromic oxide. The petrographic techniques were unable to show definitive results as to the composition. However, through the use of reference materials combined with an experienced microscopist the composition of the stone can be identified. The source of the stone could be either chromium metallic contaminant or unmelted colorant.

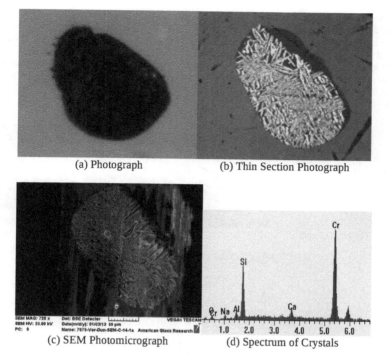

(a) Photograph   (b) Thin Section Photograph

(c) SEM Photomicrograph   (d) Spectrum of Crystals

Figure 6. Case Study #2 Stone Sample #24

Conclusion for Case Study #2

SEM/EDX analysis of stone samples #11 and #24 was able to identify the composition as iron chromite (#11) and chromic oxide (#24), respectively. Petrographic techniques were unable to determine the compositions. However, the physical characteristics and through consultation of various reference materials, a trained microscopist could identify these stones.

Case Study #3 - Stone Sample #7

The stone appeared yellowish/brown in amber glass as shown in Figure 7(a). There was a small solution sac indicating melting into the surrounding glass and small rounded nodules protruding slightly along the perimeter of the stone. Petrographic analysis revealed the presence of individual nodules, as shown in Figures 7(b). The characteristic shape and the high birefringence indicated that the nodules were likely zirconia.

SEM analysis revealed irregularly shaped nodules, as shown in Figure 7(c). EDX showed that the nodules were composed of zirconia. Between the nodules, the composition was an alumina rich glassy matrix (Figures 7(d) and 7(e)). The SEM/EDX analysis showed definitively that the nodules were zirconia. In addition, the areas between the nodules were alumina rich. The petrographic analysis could only suggest the presence of zirconia based on the shape of the crystals and their high birefringence. Since there was no crystalline structure associated with the alumina rich background, petrographic analysis was unable determine the difference in composition of the solution sac as compared to the glass. Based on the presence of zirconia nodules and background alumina, it was concluded that the stone was likely erosion of AZS refractories in the furnace.

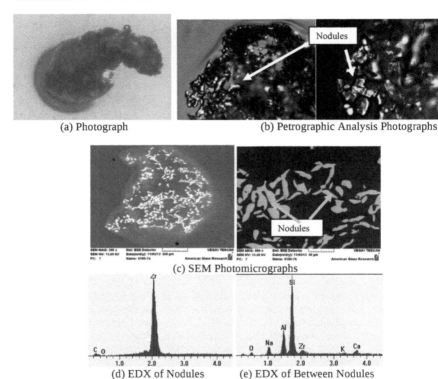

Figure 7. Case Study #3 Stone Sample #7

Case Study #3 - Stone Sample #8

The stone was a glassy knot containing white dendritic crystals, as shown in Figure 8(a). Petrographic analysis identified the stone through physical characteristics and the birefringence level as recrystallized zirconia dendrites in a viscous sac, as shown in Figure 8(b).

SEM analysis showed a view of the dendritic crystals (Figure 8(c)). EDX analysis revealed the crystals were composed of zirconia in a background matrix of highly aluminous glass, as shown in Figures 8(d) and 8(e). SEM/EDX analysis and petrographic techniques both determined the stone was composed of zirconia dendrites. However, the EDX was able to identify the presence of highly aluminous glass whereas the petrographic techniques were unable to make this discernment. Based on the zirconia dendrites and alumina background, the stone was likely the result of AZS erosion from above the flux line.

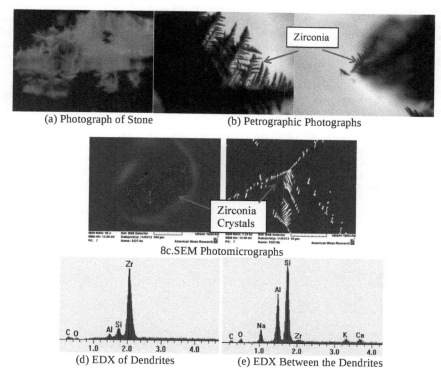

(a) Photograph of Stone   (b) Petrographic Photographs

8c. SEM Photomicrographs

(d) EDX of Dendrites   (e) EDX Between the Dendrites

Figure 8. Case Study #3 Stone Sample #8

Conclusion for Case Study #3

Both SEM/EDX and petrographic techniques were able to identify the secondary zirconia in the stones. In addition, in both stones, EDX was also able to identify the composition of the highly aluminous background material around the crystals. Not all refractories that contain zirconium also have alumina in their matrix. Therefore being able to distinguish the presence of alumina provided additional information that was used to definitively identify the source of the stones.

Case Study #4 - Stone Sample #14

The stone was white in color with seeds at the perimeter, surface cracking and a solution sac, as shown in Figure 9(a). Petrographic analysis revealed a solid, brown colored center (in polarized transmitted light). In addition, near the perimeter, there were small hexagonal crystals, lath-like crystals, and needle shaped crystals. The outer boundary of the stone also exhibited feathery type crystals, as shown in Figure 9(b). Based only on the brown coloration, this stone has typically been identified in the reference literature as a clay material. The lath-like crystals showed parallel extinction and positive elongation. Based on the crystal shape and the optical properties it was concluded the crystals were corundum. The shape of the hexagonal crystals also indicated they were corundum. The small, fine needles within the same region also showed parallel extinction and positive elongation. These needles were thinner and much shorter than the laths and thus, the combination of their optical properties and their physical appearance indicated these needles were mullite. The perimeter feather-like crystals were likely nepheline as determined by their physical nature and location.

SEM analysis revealed a large mottled area surrounding a solid center, with laths and needles near the perimeter, as shown in Figure 9(c). EDX analysis revealed the solid center was composed primarily of silica and alumina(9(d)); the mottled area was composed of silica, alumina, and soda (9(e)); the elongated lath-like crystals which were composed primarily of alumina (9(f)); and the needles were composed primarily of alumina with lesser amounts of silica (9(g)). SEM/EDX analysis provided the same information as the petrographic techniques. In addition, it provided the elemental composition of the center, laths, needles, and perimeter of the stone. The stone was a clay refractory material with some conversion to corundum, mullite, and nepheline.

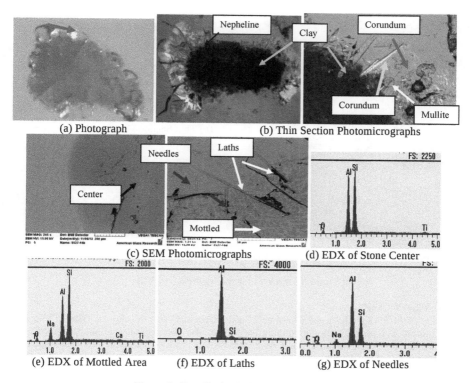

(a) Photograph · (b) Thin Section Photomicrographs · (c) SEM Photomicrographs · (d) EDX of Stone Center · (e) EDX of Mottled Area · (f) EDX of Laths · (g) EDX of Needles

Figure 9. Case Study #4 Stone Sample #14

Case Study #4 - Stone Sample #15

The stone was white in color, exhibited a large solution sac, and had many associated seeds, as shown in Figure 10(a). Petrographic analysis (Figure 10(b)) showed the stone was brown in color when examined in polarized transmitted light indicating that it was likely a clay material. Crystal grains at the perimeter of the stone had irregular shapes and moderate birefringence. Based on these properties it was concluded that these grains were corundum.

SEM analysis revealed the stone had seeds at the interior as well as the exterior. In addition there were small grains at the interior, as shown in Figure 10(c). EDX analysis revealed the small grains at the interior were composed of silica (10(d)). The remaining area of the interior of the stone was composed primarily of silica with a lesser amount of alumina (10(e)). Both the SEM/EDX and the petrographic concluded the primary stone was an alumino silicate clay. EDX identified the grains at the perimeter as being composed of silica; however, petrographically they were incorrectly identified as corundum. The stone source was likely a ceramic cullet contaminant such as a tableware item.

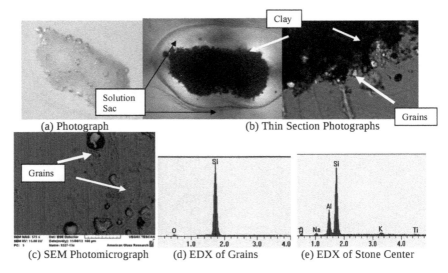

(a) Photograph   (b) Thin Section Photographs

(c) SEM Photomicrograph   (d) EDX of Grains   (e) EDX of Stone Center

Figure 10. Case Study #4 Stone Sample #15

Case Study #4 Conclusion

The SEM/EDX analysis and petrographic techniques identified the clay (alumino-silicate) in both sample #14 and #15. In determining the source, clay material can be used as a refractory material in glass manufacturing and as a ceramic such as tableware (cullet contaminant). Typically, a higher amount of alumina increases the temperature resistance of clay, thus refractories, by their nature would be composed of more alumina than a common ceramic.

SEM/EDX analysis was able to resolve the ratio of the alumina and silica in the primary portion of the stones. Thus, sample #14 with approximartely 50% alumina, was likely a refractory material. Sample #15 with a lower alumina level of 20% was likely a ceramic cullet contaminant. Petrographic analysis techniques are unable to resolve ratios of oxides and thus, cannot definitively indicate the source of the stones.

CONCLUSIONS

The current study shows that SEM/EDX analysis of stones provides comparable results to the petrographic techniques. In addition, EDX provides the analyst with the elemental composition of the stone whereas petrographic techniques are unable to distinguish certain compositional differences. EDX also has the ability to identify portions of a stone matrix that are non-crystalline which provides additional information that is also not available using petrographic techniques.

If an SEM/EDX is accessible, an experienced analyst can employ the equipment knowing it will provide at least comparable information to petrographic techniques, and in certain cases it will provide additional information that will assist in identification of the source of the stone.

CAUTIONS
1. It is imperative that proper stone preparation techniques are employed in order to identify all the components of the stones using either analysis methodology.
2. An understanding of crystal morphology and the interaction of the stones with the temperatures in a furnace can greatly assist in the determination of the source of stones.
3. Experience in both petrographic and SEM/EDX experimental techniques will increase the ability of the micrscopist to more readily identify key details of stone morphology.

ACKNOWLEDGEMENT
Ron Plenzler for petrographic analysis performed on the stones for this study.

REFERENCES:
1. Dimmick Henry, Nichols Neal, Smay Gary, "Analysis of Cord and Stones in Glass." 71st Conference on Glass Problems. Charles H. Drummond III. Columbus: A. John Wiley & Sons Inc, 2011. 81-86. Print
2. Stones and Cord in Glass, C. Clark-Monks and J. M. Parker, Society of Glass Technology, 1980
3. Guide to Refractory and Glass Reactions, Edward R. Begley, Calners Book Division, 1970
4. M. A. Knight, "Identification of Stones in Glass" The Glass Industry February , 1947
5. Environmental Scanning Electron Microscopy, Philips Electron Optics,1996
6. Energy-Dispersive X-Ray Microanalysis, Kevex Instruments , Inc., 1989

Forming

MULTI GOB WEIGHT PRODUCTION

Xu Ding and Jonathan Simon
Butcher Emhart Glass R&D enter
Windsor, CT 06095, USA

Angelo Dinitto
Butcher Emhart Glass
Savona, Italy

Andreas Helfenstein
Butcher Emhart Glass
Cham, Switzerland

ABSTRACT
The capability of producing multiple containers with different weights on the same machine line can give the glass plant more flexibility to organize production jobs, save cost on mold equipment, and reduce production loss due to job changes. To meet this need, Bucher Emhart glass has developed the multi gob weight control system, including multi weight feeder and the knowledge for how it can be applied. This paper presents the recent development of a model based setup algorithm for the Multi Gob Weight System. The algorithm makes the multi gob weight set up procedure quick, accurate, and easy for the end user. Plant trial results are included to validate the effectiveness of the setup algorithm implementation.

INTRODUCTION
The capability to produce multiple containers with different weights on the same machine line can give the glass plant greater flexibility to organize production, save cost on mold equipment, and reduce production loss due to job changes. By correctly setting up feeder control parameters, multi gob weights can be produced in a sequence to feed the corresponding forming sections and produce different containers from a single machine line. Bucher Emhart Glass has developed the Multi Gob Weight System for this purpose. The system includes the FlexIS Multi Gob Weight Feeder control system and the MGA (Multi Gob Application), a simple to use, model based software tool for determining the required setup parameters.

DEVELOPMENT
The glass gob is formed from the glass feeder device that includes a bowl shaped structure, a rotating tube, and a plunger mechanism, as shown in Figure 1, which displays the results of a feeder model used to simulate glass gob forming. The plunger is driven by a motor and can move up and down. The glass stream from the feeder outlet is cut into individual glass gob by a shearing system. The formed gob characteristics, gob weight/length/shape, are a function of the glass viscosity and the feeder control parameters, which include the tube height, the plunger motion (plunger position, plunger moving distance, and plunger motion phase), the shear control, the feeder tube motion, and position, and glass viscosity. All these feeder control parameters can influence the gob weight, length and gob shape. When one parameter is changed for one gob, it may have significant effect on adjacent gobs too. The relationship between gob weight and feeder parameters can be expressed mathematically as:

$W =$ f(plunger height/stoke/cam profile/shear control, successive plunger

height/stroke/cam profile/shear control, feeder tube motion, temperature)

where W represents the formed gob weight.

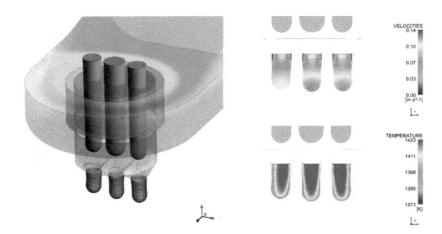

Figure 1. A glass feeder model showing glass gob forming

In Bucher Emhart Glass, just such a mathematical model has been developed to correlate the major feeder control parameters and gob weight as well as the influence on adjacent gobs. The model can establish the relationship among the gob weight and feeder control parameters such as the plunger height, stroke, phase, shear differential etc. The model can predict the gob weights based upon the feeder control parameters input. The model can also generate required feeder control parameters to produce desired gob weights and cutting pattern. For a new feeder configuration or dramatically changed feeder configuration, a calibration step is required for the mathematical model.

Several field tests have been completed, which validated the model prediction accuracy. Figure 2 shows the validation test results from BEG Research Center for an NIS machine producing 2 gob weights, 3 gob weights, and 4 gob weights. As shown, the model can setup the production pattern correctly and all measured gob weights match the set points.

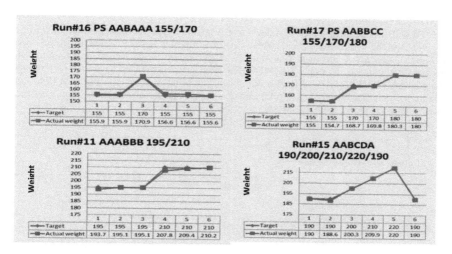

Figure 2. Model validation test

Figure 3 shows the results from a 4 section NIS machine with two gob weights TG (Triple Gob) NNPB production setup using the model. The model was used to setup the gobs for the inner, middle and outer cavities following the same pattern. The outer gob showed more discrepancy than the middle and inner gobs. With feeder needle adjustment, the discrepancy could be reduced within tolerance for the production.

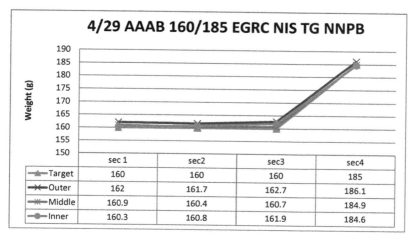

Figure 3. TG, NNPB, 4 section NIS machine production with two gob weights

STANDALONE APPLICATION

BEG (Bucher Emhart Glass) has developed a software application program to provide the multi gob weight setup application, which will be referred to as the MGA (Multi Gob Application). The MGA has been developed to make the multi gob weight production setup more practical and user friendly. The MGA has been implemented as a standalone program running on the FlexIS Universal Console Computer. The program communicates with FlexIS to obtain the current glass machine setup information such as feeder parameters, firing order etc. Figure 4 shows the workflow of the MGA.

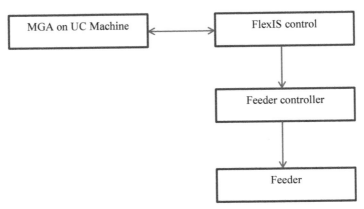

Figure 4. The MGA workflow chart.

The main interface of the standalone application is shown in Figure 5. On the MGA user interface, the user can enter the desired gob weights for each section. The program generates the feeder parameters and loads them to the FlexIS control to realize the desired multiple gob weight production. After setup, the user can measure the actual gob weights and fine-tune the feeder control for even more accurate weights through the program interface.

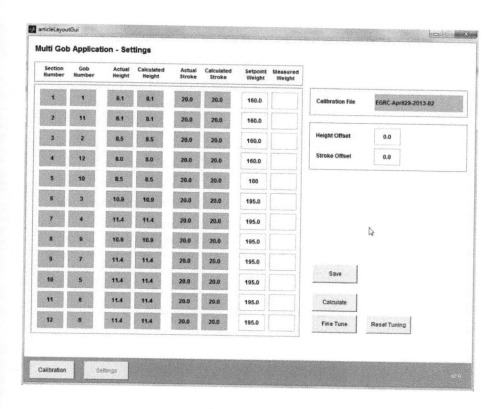

Figure 5. The main interface of the standalone program

TEST RESULTS

The MGA was tested in field plant. Two test result samples are summarized in this section. Figure 6 shows the test results on a customer's machine, for an 8 section, single gob, 3 gob weight production. The target gob weights are 430gram, 450 gram, and 460gram respectively. After the MGA setup, about 5gram~6.5gram gob weight discrepancy was observed. A fine tune step was performed. After the first fine tune, the actual container weights are within 1.6 gram of the set points. All gob weights were within the gob weight variation tolerance for this particular production.

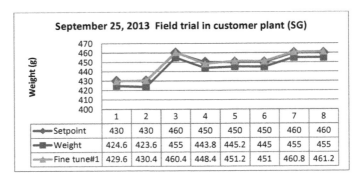

	1	2	3	4	5	6	7	8
Setpoint	430	430	460	450	450	450	460	460
Weight	424.6	423.6	455	443.8	445.2	445	455	455
Fine tune#1	429.6	430.4	460.4	448.4	451.2	451	460.8	461.2

Figure 6. The test results on a customer machine (SG, 8 sections, three gob weights)

Figure 7 shows the field results on a second customer machine, which is a 12 section, double gob, 2 gobs weight production. The target gob weights are 933gram and 952gram. After the MGA setup, about 5~6gram gob weight discrepancy showed up on some of the sections. A fine tune step was performed. After first fine tune, the actual container weights are within 3 gram to the set points. All gob weights are within the gob weight variation tolerance for the production.

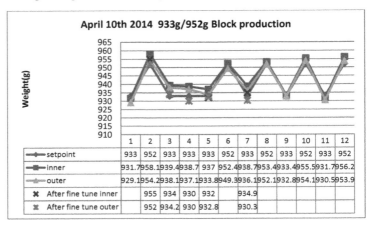

	1	2	3	4	5	6	7	8	9	10	11	12
setpoint	933	952	933	933	933	952	933	952	933	952	933	952
inner	931.7	958.1	939.4	938.7	937	952.4	938.7	953.4	933.4	955.5	931.7	956.2
outer	929.1	954.2	938.1	937.1	933.8	949.3	936.1	952.1	932.8	954.1	930.5	953.9
After fine tune inner		955	934	930	932		934.9					
After fine tune outer		952	934.2	930	932.8		930.3					

Figure 7. Field test results with a second customer (DG, 12 sections, 2 gob weights)

SUMMARY

A mathematic model was developed to establish the correlation between gob weight and gob feeder control parameters. By using the model, the feeder control parameters can be determined to achieve the desired gob weight for each individual machine section, for a multi gob weight production. The mathematical model is implemented in a standalone software application called the MGA (Multi Gob Application). The software communicates with the Bucher Emhart Glass FlexIS control system to provide the feeder control parameter setup for multi gob weight production. The field trial results of the multi gob weight application (MGA) validated the mathematical model predictions and demonstrated that the software provides a practical and user-friendly tool to setup multi gob weight production. The MGA product will be available soon in the market.

CLOSED LOOP CONTROL OF GLASS CONTAINER FORMING

Jonathan Simon
Butcher Emhart Glass R&D enter
Windsor, CT 06095, USA

Andreas Helfenstein
Controls Development Engineer
Emhart Glass SA
Cham Switzerland

ABSTRACT

Recently developed closed loop control systems offer a new means to help container glass manufacturers meet the ever increasing industry and customer expectations for improved yield and quality. In these closed loop systems, the process is automatically adjusted based upon actual measured values. Such closed loop controls, have now been developed and introduced into commercial production in two key areas: 1) Plunger Up Control and 2) Blank Cooling Control, in which the rise/dwell time of the plunger and the blank mold temperatures, respectively, are automatically controlled. The technical development of these new control systems, the challenges that needed to be met, and the experience of glass manufacturers adopting these systems will be presented.

INTRODUCTION – THE NEED FOR CLOSED LOOP CONTROL

The transformation of a molten glass gob into a glass container requires a close coordination between heat removal and the progression of the mechanical deformation of the glass. The necessity for this close coordination is due to the exponential dependency of the glass viscosity on the temperature [1], and thereby the thermal state of the glass. If, for example, excessive heat removal takes place prior to some forming step, such as the pressing of the parison, the glass will be too viscous, and it may take excessive pressing force to complete the stroke of the plunger. Variation in heat removal in the blank molds, will influence the rate at which the parison, runs (stretches under the influence of gravity) prior to being inflated in the blow molds, which will result in an incorrect distribution of the glass thickness. If the total heat removal is insufficient, the container will reheat on the conveyor and begin to lean.

It can therefore be seen that any external disturbance which affects the initial temperature of the glass, the viscosity at a given temperature (glass composition), heat removal rates, or the mechanical motion of the forming tooling (e.g. plungers), or swabbing technique, will result in undesirable variation in the finished product. As depicted schematically in Figure 1, such variation, if left uncontrolled, results in out of specification or defective containers. These must be rejected reducing the amount of shippable product (pack rate). The end result of such variability and resulting, loss will be a reduction in the profitability of the operation.

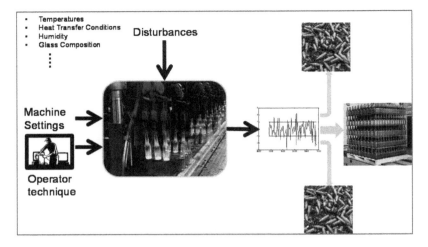

Figure 1 Effect of Disturbances on Pack Rate (schematic depiction)

As the acceptance specifications imposed by the container manufacturer's customers become stricter, the sensitivity of the pack rate to disturbances increases. It can be seen that to reverse this trend, and improve or restore profitability, some form of corrective action is required to reduce the impact of the disturbances on the process. Closed loop control, in which key process variables are measured, and compensating adjustments are automatically applied, provides an approach for achieving the desired reduction in variability and ultimately increasing profitability of the operation.

CHALLENGES FOR CLOSED LOOP CONTROL OF FORMING PROCESS

Closed loop process control has been widely adopted in many industries. As an exception, the glass forming process presents particular challenges, which have, until recently limited it's application to this industry. In general there are three major challenges, which must be addressed in order to obtain practical control of the glass forming process:

1. Finding suitable measurement systems to measure process variables within the harsh conditions found in a glass-forming machine.
2. Providing control strategies which take into consideration the special features and characteristics of the forming process
3. System integration to provide seamless connections between the measurements systems, the control algorithms, the machine timing control hardware, and the machine timing control user interface.

As will be described in the subsequent sections, these challenges have now been successfully addressed for two different parts of the forming process now allowing the benefits of closed loop control to be obtained for pressing (Plunger Up Control), and blank side thermal management (Blank Cooling Control).

CLOSED LOOP CONTROL OF PARISON FORMING - PLUNGER UP CONTROL

Background
     Two main types of processes are commercially utilized for forming glass containers. They are distinguished by the means used to form the hollow interior of the parison (perform). In the Blow and Blow process, the hollow interior is created using compressed air (with a mechanical plunger only used to make a small starting dimple in the glass). In the second process type, known as Press and Blow, a mechanical plunger is pressed into the gob to form the hollow interior. The Press and Blow process can be further subdivided into NNPB (Narrow Neck Press and Blow), and WMPB (Wide Mouth Press and Blow) depending upon the type of container (bottle or jar respectively) that is formed. The positive displacement nature of the Press and Blow process provides parisons with more consistent (compared to Blow and Blow) wall thickness. All else being equal, this ultimately produces containers with reduced variability. These benefits, come however with a "cost" in terms of the sensitivity of the process to other disturbances. As will be seen, it is therefore advantageous to provide a closed loop system in order to obtain the full benefits of the Press and Blow Process.

Parison forming process (Press and Blow or Narrow Neck Press and Blow)
     To get a better understanding of the potential sensitivity of the press and blow process to disturbances and how this can be mitigated by a closed loop control, it is helpful to first look at the process in further detail. A view of the interior of a blank mold, as the plunger forms the parison, is shown in Figure 2. As shown (right panel) the plunger is driven by a pneumatic mechanism. With this arrangement, the rate of travel of the plunger is dictated by a force balance between the pneumatic force applied by the piston and the resistance of the molten glass to being deformed and displaced by the plunger. As a result, any changes in either the mechanical/friction situation of the piston, or the viscosity distribution of the glass will result in changes to the plunger motion profile (position as a function of time). As the partially formed parison is continuously losing heat to the blank molds and plunger, changes in motion profile become coupled with the temperature dependent viscosity of the glass. This can result in a number of issues and possible defects. In the extreme case, the plunger may come to rest before the finish (opening) of the container is fully formed as depicted in Figure 3, clearly an unacceptable situation.
     A typical motion profile for a plunger is shown in Figure 4. As shown, the overall motion plunger motion may be divided into distinct phases:

- Plunger moves to loading position ready to receive next gob
- Initial plunger up motion as plunger rises quickly to meet the gob
- Glass distribution as the plunger deforms and displaces the gob forming into a parison
- Plunger dwell where the plunger holds the shape of the parison as it cools (and contracts) until it becomes stable enough to invert into the blow molds.
- Plunger down where the plunger is retracted, freeing the parison to be inverted.

     The plunger rise time is defined as the time from the start of the initial plunger up phase until the start of the plunger dwell. One quantitative measure of the variability in the pressing process is then the variation in the plunger rise time. A typical actual situation for an uncontrolled (no closed loop control) plunger is plotted in Figure 5, which shows the variation over time of the plunger rise time for a particular cavity. The periodic decrease in rise time, which occurs at approximately 300 cycle intervals, is particularly notable. These sharp decreases correspond to the operators applying swab dope to the blank molds. It is hypothesized that the effect of the swabbing is to change the friction situation between the glass and the blank molds providing lubrication, which allows the parison to be formed more quickly. Whenever the plunger rise time is increased, the plunger dwell time is decreased (for fixed start of plunger down). Thus not only is the pressing time varying, the amount of time for the parison to become stabilized (plunger dwell) is varying too. It is clear that there is a potential to reduce this variability, if it could be measured and controlled.

Figure 2. Plunger forming the parison

Figure 3. Incomplete travel of plunger resulting in unfilled finish

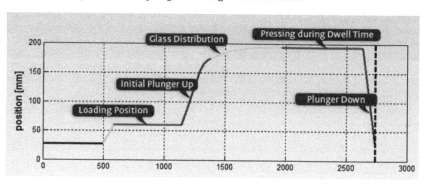

Figure 4. Typical plunger motion profile

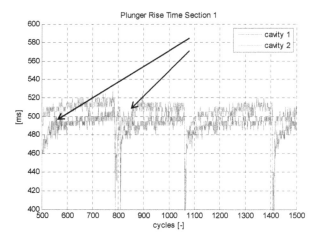

Figure 5. Variation in plunger rise time, showing effect of swabbing

Measurement System

Obtaining a good measurement of the process to be controlled is a prerequisite for implementing a closed loop control system. For the case of the plunger up control system we have been able to utilize the Emhart PPC (Plunger Process Control) system to provide the necessary measurements as shown in Figure 6. In this system, the actual measurements are made by the full stroke plunger sensor, as shown in Figure 6, lower right. This provides a variable capacitance, which can be detected without any wired connection to the moving plunger, avoiding problems with flexing and breaking connections, or sliding contacts. In addition the design is rugged enough and proper materials are utilized to allow it to withstand elevated temperatures experienced within the plunger housing.

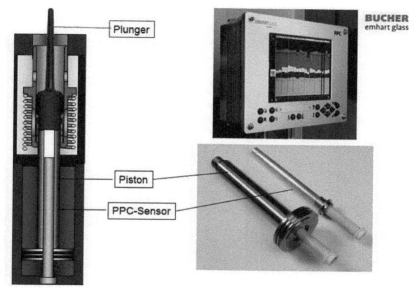

Figure 6. Emhart PPC system

Closed Loop Control Strategy

The Plunger Up Control controls utilizes an FPS (Flexible Pressure System) control valve which can be set to supply four different pressure values (three different positive values and zero) to the plunger's pneumatic cylinder over the duration of the pressing cycle. As depicted in Figure 7, typically we have:

- The highest pressure, P1, is used to quickly move the plunger into contact with the glass (initial plunger up motion)
- The pressure is reduced to an intermediate value P2, to actually press (distribute) the glass
- The pressure is reduced further to P3 to maintain intimate contact between the glass and the tooling as the parison stabilizes (plunger contact)

The closed loop control then automatically adjust the pressure level P1 and P2 (maintaining them in a fixed ratio) to keep the measured rise time as constant as possible.

Based upon the measured press curves, the closed loop control also automatically adjusts the time at which the pressure level switching occurs, so that the pressures are applied at the appropriate phases in the press. User configurable parameters allow the ratio of P1 to P2 to be adjusted and also allow the user to specify the desired switching times (as a percentage of each motion phase).

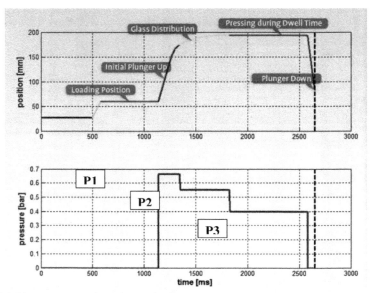

Figure 7. Multi-pressure pressing using FPS valves

System Integration

In order to actually realize the control strategy in a practical implementation we must bring together the measurements, the timing system, the control algorithm, and the user interface. As shown schematically in Figure 8, a software architecture for this purpose has been developed which allows all of the necessary pieces to communicate and exchange data. As shown, the central piece of this architecture is a software module, which is called the Flexternal. This provides the necessary communication bridge between the components, allowing them to communicate and exchange data.

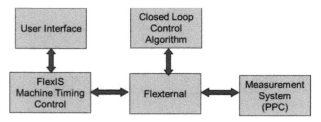

Figure 8. System integration architecture

A key goal in developing the closed loop applications was to make it as simple as possible for the plant personnel to use and configure them. To achieve this, the user interface is provided simply by extending the existing control system user interface with the addition of a small number of screens. Each new screen preserves the look and feel of the existing control, and has been made as streamlined as possible to provide just the essential functionality. An illustrative view of the user interface for the plunger up control is shown in Figure 9. As shown, the desired (setpoint) value for the rise time can be set for each section and cavity of the machine. The user interface displays the setpoint value and actual value for each cavity. Individual control loops can also be turned on and off from this screen. Colors are changed in the display of the actual values to indicate control loops that are currently at their user-configured limits.

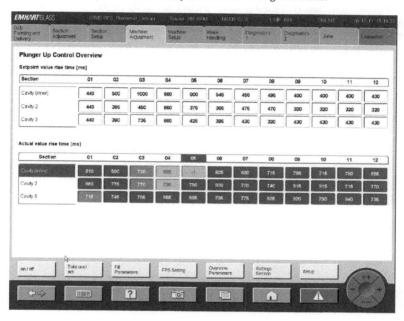

Figure 9. Plunger up control user interface overview screen

Results

The closed loop plunger PUC (Plunger Up Control) has now been installed and successfully utilized at numerous glass container production facilities. Figure 10 shows field trial results from one section of an actual machine in the field, illustrating how the closed plunger dwell times are driven smoothly to the desired setpoint value (left hand panel) after the control is turned on. It is notable that the pressure level (P2) required to achieve these dwell times (right hand panel) is different for each cavity. This could be due, for example, to differences in the friction situation for the different cavities. To arrive at a uniform dwell time for all of the cavities without a closed loop would be quite difficult. Even if an operator had a measurement system to view the dwell times, he/she would have to make the adjustments through a trial and error process. Depending upon the skill of the operator this could be a quite lengthy procedure even for just a single section. It would be of course much longer for a full (e.g. 12 section machine).

A comparison of open loop and closed loop PUC for multiple machine sections is shown in, Figure 11, which plots data for all three cavities of Section 7 through 12 (labels S7,S8... S12) of a twelve section triple gob (3 cavity) machine. Considerable variation in rise time is evident while in open loop (prior to cycle 1200). It can be seen that this variation is dramatically reduced once the closed loop control is switched on. It is particularly interesting to observe the behavior of Section 10 (S10). Prior to switching the closed loop on, it is apparent that the plunger rise times are often excessive, indicating that the plunger may be getting stuck (due to excessive friction) and then finally breaking away during its initial travel. The closed loop control evidently corrects this situation, by applying additional pressing pressure in the early phases, overcoming the friction.

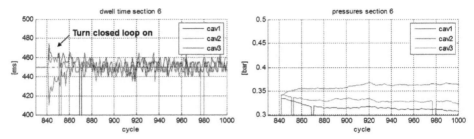

Figure 10. Plunger up control–closed loop control brings dwell time to set point value

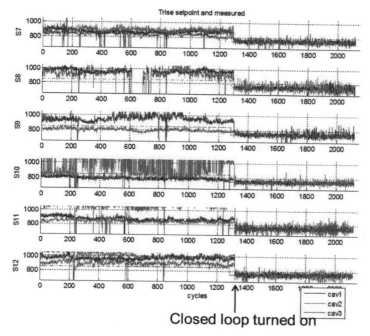

Figure 11. Comparison open and closed loop control of plunger rise time

CLOSED LOOP CONTROL OF BLANK SIDE THERMAL PROCESS – BLANK COOLING CONTROL

Background

The outer shape of a container is formed by inflating the parison within the blow molds, and is therefore relatively constant. In contrast, the inner wall location is left free and so the wall thickness distribution of the container is strongly dependent upon the thermal state of the parison when it is delivered to the blow molds. The thermal state of the parison affects how it stretches under the influence of gravity prior to being blown, and also how the glass distributes as it inflates. The thermal state of the parison is in turn dependent upon the thermal conditions in the blank molds. As these conditions are subject to numerous disturbances, there is a potential to reduce variability in the final glass distribution, by properly managing the thermal balance in the blank molds. The successful application of this concept was previously described elsewhere [2] for a prototype proof of concept test. This technology has now matured, and the following sections will describe how it has now been fully integrated within the Emhart FlexIS control system.

Blank Cooling Process

A schematic depiction of an axially cooled blank mold is shown in Figure 12. As shown, cooling air is forced through a set of axial cooling passages arranged around the circumference of each mold cavity. This cooling air is used to remove the heat, which is supplied to the mold from the glass as it is transformed into a parison. Cooling air can be arranged to flow from the top down (as shown) or bottom up. In either case the airflow is modulated by on/off valves, which open and close cyclically during each successive parison forming cycle. The fraction of the cooling cycle that the valves are held open affects the overall temperature of the molds. Analytical studies (Simon & Braden, 2012) indicate that the temperature varies both spatially through the mold and temporally over the forming cycle. Ultimately, we are interested in the heat removal from the glass, which in turn responds to the temperature distribution in the blank molds. Thus we indirectly adjust the heat removal from the glass, by increasing or decreasing the cooling duration of the molds. This process may be disturbed by a number of causes including, changes to the cooling air temperature, changes in fan supply pressure, changes in initial temperature of the glass (gob temperature), changes in the heat transfer situation between the glass and the molds (particularly swabbing). Closed loop control of this process has been introduced as a means to automatically adjust the airflow to counteract the effects of these disturbances.

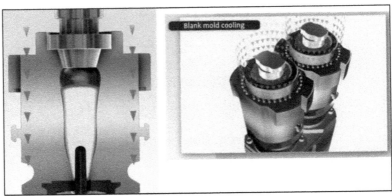

Figure 12. Axial cooling of blank molds

Measurement System

The TCS, a rail mounted, movable pyrometer, Figure 13, is utilized to obtain measurements of the blank mold glass contact surfaces. As described in further detail in (Simon & Braden, 2012) the pyrometer traverses along a horizontal rail and can rotate in the vertical and horizontal plane allowing it to measure temperatures from all the cavity halves of every each section in a machine. Purge air is supplied to the unit providing cooling and keeping the optics clean, which allows it to withstand the harsh glass making environment.

Figure 13 TCS 3-Axis movable pyrometer provides blank mold temperature measurements

Closed Loop Control Strategy

The basic control strategy chosen is to adjust the cooling duration (within each cycle) based upon the measured values (provided by the TCS). In order to make the system usable in as many situations as possible, flexibility has been provided to allow the user to configure numerous variants of this basic strategy. Regarding the cooling duration two cases can be handled:

- The time at which the cooling is turned on is adjusted and the time it turns off is held constant
- The time at which the cooling is turned off is adjusted and the time that it turns on is held constant

The following situations can be handled regarding number of measurements and actuators (valves)

- Individual valve for each cavity mold half
- Single control valve for multiple cavity mold halves
- Measurement of every cavity mold half
- Single cavity measurement for multiple cavity mold halves

The solution that is employed for each possible combination of number of measurements and number actuators is shown in Table 1. It is also important that the user can configure limits on the control, so that cooling only occurs during a portion of the cycle that is acceptable to the user. To handle this, user configurable limits, can be imposed on the influenced event (either the time that the cooling is turned on or the time that it is turned off). Further, the control algorithm is able to handle the situation where the output is limited so that it does not experience any windup phenomena.

Table 1 Blank Cooling Control Configurations

n = number of cavities		Number of Valves	
		n	1
Number of Measurements	N	Individual Loop For Each Mold Cavity Half	Average Measurements
	1	Slave valves to move together	Individual Loop for Each Section Half

Control loop tuning was another important aspect that needed to be addressed to make the system widely applicable without requiring extensive testing or expertise on the part of the user. In particular, the dynamic characteristics of the process as seen by the measurement system depend upon the number of cycles between measurements, which is determined by the round trip time of the TCS. This depends upon the number of measurements made within a round trip, and therefore on the number of cavities and sections, which is both job and machine dependent. To address this issue, an analytical method was developed and implemented to compute controller gains explicitly accounting for the actual TCS round trip time.

System Integration

The general Flexternal architecture described previously for the PUC and shown in, Figure 8, is also used for the Blank Cooling Control, where in this case the TCS provides the measurements. The user interface is an important component of the overall system. The design which has been developed is illustrated in, Figure 14, which shows the screen used for making adjustments to a particular section. This screen illustrates how the key capabilities needed for a practical control are provide: As can be seen, the screen provides the ability to:

- Set the desired setpoint temperatures, compare them with the actual values
- Turn the control on and off.
- Set upper and lower limits on the influenced event (expressed cyclically as a 0-360 machine degree setting)
- View the current cooling duration (cooling off minus cooling on) in machine degrees

Additional screens are available for configuring the site-specific aspects of the control, and to provide an overview of the process. As for the PUC, the user interface has been fully integrated into the overall FlexIS timing system user interface, preserving the same look and feel as the other screens. This is encourages acceptance of the new system as it allows the operators to move comfortably to the closed loop controls without having to learn how to use a new interface.

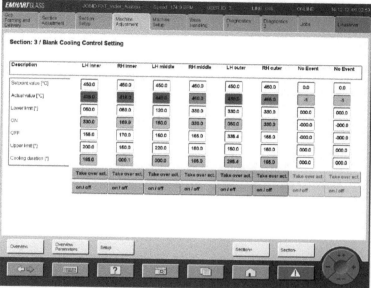

Figure 14. Blank cooling control user interface

RESULTS

In general, it has been found at numerous installations of the Blank Cooling Control System that temperatures can be controlled quite well, with a significant improvement in the variation of the temperatures across the machine and over time. This is well illustrated by a typical example shown in Figure 15, where the closed loop control is turned on at approximately 9:00. It can be seen that the temperatures then converge to the setpoint value (470 deg C) and typically stay within +/- 5 deg range of the setpoint (a few outliers occur where larger process upsets may have occurred, for example a jam up).

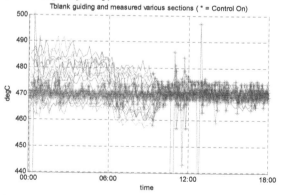

Figure 15. Switching blank cooling control on for sections 2 to 6

The ability of the closed loop to handle large upsets is illustrated in, Figure 16, where the reaction to an overheated blank mold is recorded. (The cold mold cycle may have been inadvertently left on after a section stop causing the mold to overheat). As can be seen, there is a large initial reaction of the controller, which in fact drives the output to the limit value. As desired, the control then remains at the limit, and finally begins to lift off as the mold continues to cool. Finally the mold temperature and controller output returns to normal, controlling about the setpoint. Further examples of utilizing the closed loop blank cooling control are provided in (Simon & Braden, 2012).

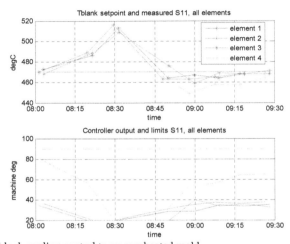

Figure 16. Reaction of blank cooling control to an overheated mold

OVERALL PERFORMANCE
Studies have just begun to further quantify the overall benefit of applying the closed loop control to the forming process. A preliminary indication of the desirable effects of the closed loop control is given from the inspection results shown in, Figure 17, which plots the losses due to wall thickness deviations before and after switch on the closed loop controls (in this case both the Plunger Up Control and the Blank Cooling Control were turned on at the same time). A clear reduction in the losses can be observed in the approximately 9 hours after the closed loops are activated.

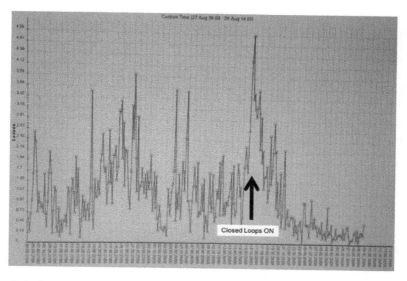

Figure 17. Reduction in losses due to wall thickness after switching on closed loops

SUMMARY AND CONCLUSIONS

When forming glass containers, unavoidable disturbances affecting either the mechanical forming or the heat removal process can result in undesirable variability and defects in the final product. Closed loop control technology can help compensate for such disturbances, potentially reducing variability and increasing overall profitability of the commercial production process. Two different applications of closed loop control, Plunger Up Control and Blank Cooling Control have now been successfully developed and integrated into the existing FlexIS timing control system making them available for general use. The development of these systems, and examples of the actual field results of using these systems in operating glass plants has been presented. These results show that overall the closed loop controls provide the capability to set the process to a desired setpoint value and then hold it there with a reduction in variability compared to the open loop case. Preliminary indications from a study looking at the overall effects of using the closed loop control indicate a clear reduction in the overall losses once the closed loop controls are turned on. These encouraging results will motivate us to continue the development of closed loop controls, which we expect to further improve the glass forming process.

REFERENCES

1. Tooley, F. V. (1984). *The Handbook of Glass Manufacture Vol II.* New York: Ashlee Publishing.
2. Simon, J., & Braden, M. (2012). Closed Loop Control of Blank Mold Temperatures. *72nd Conference on Glass Problems: Ceramic Engineering and Science Proceedings, Volume 33.* John Wiley and Sons.

HARD GLASS - COMMERCIAL PROGRESS OF THERMALLY STRENGTHENED CONTAINER GLASS

Ken Bratton, Steven Brown and Tim Ringuette
Bucher Emhart Glass
Windsor, CT 06095, USA

Dubravko Stuhne
Vetroconsult Ltd.
Bülach, Switzerland

ABSTRACT

A developmental program and market introduction is underway between Bucher Emhart Glass and Vetropack Austria. In this initiative, a new strengthening machine has been installed and is running in Vetropack's glass plant located in Pöchlarn, Austria. The market introduction is based on a 330 mL beer bottle being produced for a local Austrian brewer that is a light-weighted version (200 g) of an existing returnable bottle (300 g) – a weight reduction of 33%. This paper will present the results of the initial filling line trial together with some lessons learned along the way. It will also include the results of a study performed for Bucher Emhart Glass and Vetropack by Stazione Sperimentale del Vetro (SSV) located in Murano, Italy, regarding the differences between annealed and heat strengthened glass in terms of the resistance of the glass to impact and handling-induced defects.

INTRODUCTION

Bucher Emhart Glass has produced a glass container tempering machine which takes inspected, annealed, room temperature ware and subjects the containers to a heating cycle followed by a rapid quenching cycle to impart beneficial compression layers over all surfaces. The machine was installed in Vetropack's Pöchlarn, Austria facility in the summer of 2013 and is currently treating 330 ml returnable beer bottles weighing 200 g to replace the current 300 g returnable bottle with equal or better performance results.

The strengthening treatment consists of the following steps listed in chronological order: depalletizing, conveying, laser etch for identification purposes, stacking, heating, cooling, cold end coat and inspection. In addition, a QC sampling line is installed to evaluate the integrity of the residual stress with the use of custom-designed sand abrasion machines (both inside and outside surfaces), residual stress inspection machines and with sampling burst and impact tests.

The goal is to produce rugged containers that are suitable for the returnable beer market and to outperform the heavier returnable annealed bottle by upwards of 30% in terms of burst and 10% in terms of impact. It is also expected that resistance to breakage due to drop, thermal shock, and vertical load will be significantly improved as will resistance to wear or scuffing.

PROCESS DETAILS – THE HEATING CYCLE

In the strengthening process, annealed containers are heated in a special gas-fired tempering lehr to a uniform temperature of 600-625°C. The rate of heating is shown in Figure 1, which was estimated using convection and radiation heat transfer analyses.

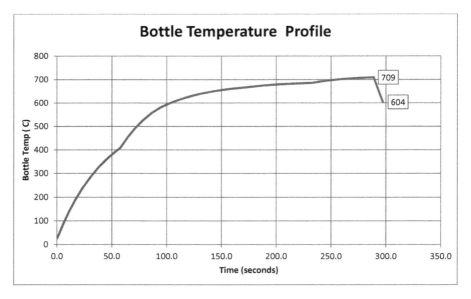

Figure 1. Bottle heating temperature profile

The trick in the heating cycle is to heat the glass as quickly as possible to avoid distortions but not so fast that the glass breaks or does not soak through the thickness and provide an even distribution. The approximate convective heat transfer coefficient in the lehr is estimated at 65 W/m^2-K. The dip in the temperature at the end of the cycle shown in Figure 1 is due to the time between exiting from the lehr and the measurement of the surface temperature. This is a critical time period as the exterior surfaces of the glass will quickly transfer heat to its surrounding via radiation cooling. This effect is shown in Figure 2.

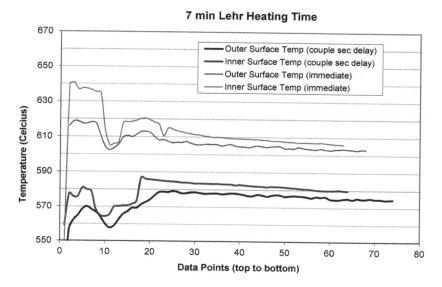

Figure 2. Bottle temperatures after exiting the lehr

Note the decrease in temperature between the data taken immediately upon exit from the lehr versus the data taken after only a few seconds - the glass has already cooled by about 30°C. Also note the difference between the inner and outer surface temperatures ranges from a few degrees to a maximum of 20°C in the finish area of the bottle. The data for Figure 2 was obtained by scanning both outside and inside surfaces of the bottle simultaneously with opposed pyrometers where the pyrometer on the inside surface utilized a prism–shaped mirror.

The rapid heat removal during the first few seconds is primarily due to radiation cooling as the glass is about 600°C hotter than its surroundings. The emissivity of glass varies with color (Green > Amber > Flint) and with temperature (approx. 0.95 between 0-200°C and 0.72-0.82 between 250°C and 1,000°C).

PROCESS DETAILS – THE COOLING CYCLE

The heated bottles are quickly picked up into the heat strengthening machine and rapidly cooled, as evenly as possible, with compressed air from three different supplies hitting the following surfaces: all interior surfaces, exterior sidewall, and the base. The flow of air is controlled through a series of nozzles and orifices to tune the velocities and the flow distribution patterns to maximize the heat removal in a uniform fashion.

Because most glass containers have widely varying wall thicknesses, one can expect the thicker sections to cool slower than the thinner sections. This must be managed so that the thicker sections do not "set-up" by going through glass transition after the thinner sections thereby imparting secondary stresses on the thinner glass. To complicate things further, thinner sections need higher rates of heat transfer to achieve enough temperature difference between the core and the skin of the glass as it cools through transition to achieve surface compression. The net result is that it is practically impossible to achieve a uniform level of compression on all the glass surfaces so the expectation must be to accept a large variation as long as there is no tensile stresses on the surfaces, no extreme levels of buried tension (levels greater than 50 MPa should be avoided) and some acceptable minimum limit on the depth of the compression layer.

A typical transient cooling curve for the various sections of a bottle is shown in Figure 3.[1] A typical plot of the resulting (non-uniform) stress distributions are shown in Figure 4-6. The stress values used to create these plots were measured in an immersion polariscope. From these measured stress maps, and knowing the wall thickness, it is possible to estimate the heat transfer coefficients. These numbers vary over the surface of the bottle but typically range between 150 W/m^2-°K to 300 W/m^2-°K.

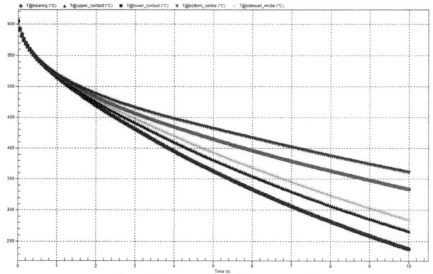

Figure 3. Temperature prediction upon cooling

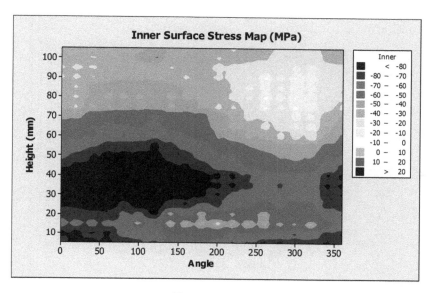

Figure 4. Inner stress map

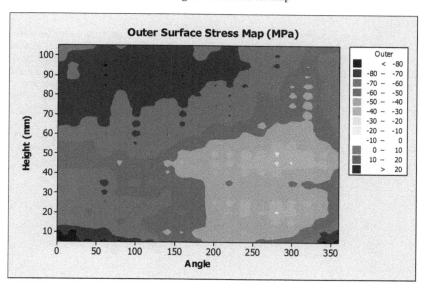

Figure 5. Outer stress map

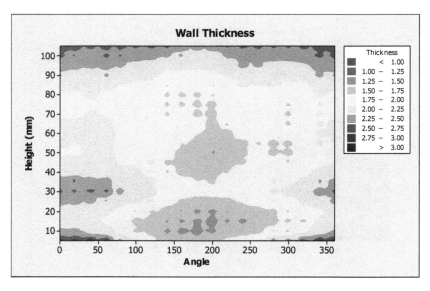

Figure 6. Wall thickness map

IMPROVED DURABILITY

The heat strengthening process offers improvements in all of the environmental loads that most containers are exposed to; loads like internal pressure, thermal shock, vertical load, drop and impact. In addition, we have discovered through cooperation with SSV that the residual compression stress layer also resists scuffing and improves the durability of the surface, which adds further benefit for the use of thermal strengthening in the returnable container markets.

In a series of tests using annealed and tempered flat plate, SSV has discovered that the surface compression resists penetration of particle impact when subjected to a controlled sand abrasion environment.[2]

From Figure 7, we can see that the average flaw depth for the annealed glass was 323 microns (165 min, 561 max) compared to only 193 microns average depth for the strengthened glass (137 min. 279 max) for an improvement of 60%.

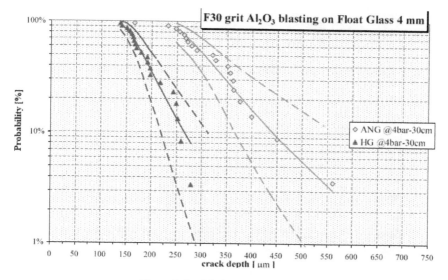

Figure 7. Particle impact test results

LOOP TEST RESULTS CONDUCTED IN THE LABORATORY

The Research and Teaching Institute for Brewing in Berlin (VLB) was tasked with subjecting approximately 800 bottles (half strengthened and half annealed) to a series of 40 loops which included an AGR line simulation followed by a caustic wash cycle. After a predetermined number of loops, samples were pulled for burst, impact and drop tests. The caustic wash consisted of an 8 minute soak in 80°C water mixed with a combination of NaOH and a cleaning agent (Sopura). The line simulator was set up to simulate a line speed of 810 bpm.

The differences in the amount of scuffing were significant with the strengthened bottles having much less damage. This is shown in Figures 8 and 9.[3]

Figure 8. Scuffing of annealed bottles

Figure 9. Scuffing of strengthened bottles

A comparison of the internal burst pressure for samples tested after loops 0 (initial), 5, 10, 15, 20, 30 and 40 is shown in Figure 10 and the impact resistance is shown in Figure 11.

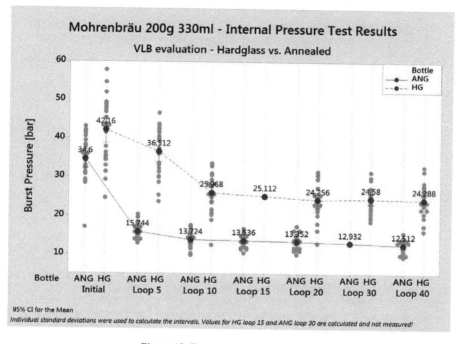

Figure 10. Burst pressure comparisons

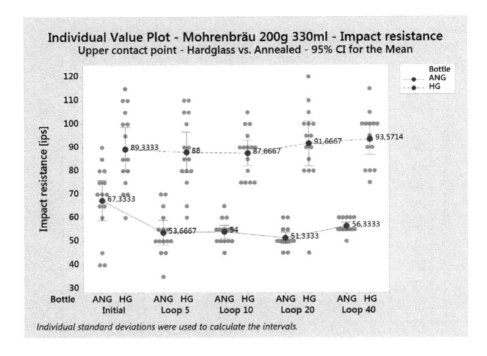

Figure 11. Impact test comparisons

ACTUAL FILLING LINE RESULTS

A filling line trial was run at a customer's site which included 25 loops through an installation that included: uncapping, crate unpacking, bottle washing (17 minutes at 81°C, alkaline bath and caustic wash), inspection, filling with carbonated water, labelling, crate packing and crown capping. After the trial, the standard 300gm returnable bottle, the annealed light-weight 200 gm bottle and the thermally-strengthened light-weight 200 gram bottle were burst and impact tested to determine the integrity of the designs after the 25 loops. The results are shown in Figures 12 - 14.

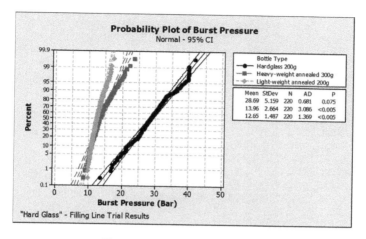

Figure 12. Burst pressure comparisons

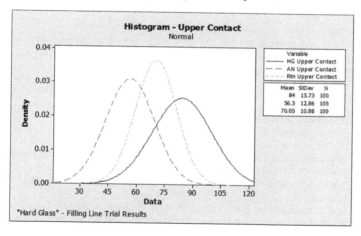

Figure 13. Upper contact impact comparisons

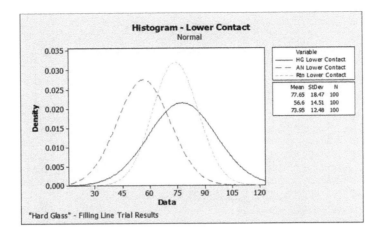

Figure 14. Lower contact impact comparisons

CONCLUSIONS

The heat strengthening process can be a viable means of reducing glass weight while maintaining (or increasing) strength. It also is a good candidate for returnable ware due to the improvements in scuff resistance and in minimizing damage penetration. On average, the weight can be reduced by about 30%, but the practical limit is reached when the production variations result in wall thicknesses below 1.5 mm because thin glass is very difficult to thermally strengthen. In fact, glass thicknesses less than 1.5 mm will not strengthen much at all and will behave, at least locally, like annealed glass.

The improvement of thermally strengthened glass over annealed glass will vary depending on the type of load (duration, magnitude and concentration). As a general rule of thumb, the improvements are greater for loading that tends to spread the load – such as thermal shock, burst and vertical load (perhaps as high as 50% improvement) but is less significant for point loads such as bottle-to-bottle impact where the improvement can sometimes be as low as 10%.

Thermally strengthened glass looks and feels identical to annealed glass but can have different breakage patterns. In general, the strengthened glass will break into smaller and blunter fragments as compared to annealed glass. On the microscopic level, we have also found that for strengthened glass, the fractured edges will actually be slightly rounded, or chamfered.

REFERENCES

1. Data courtesy of Stazione Sperimentale del Vetro (SSV – Murano, Italy), 2014.
2. "Sharp Particle Blasting Calibration – Optical Flaw Characterization on Corundum Blasted Float Glass", Stazione Sperimentale del Vetro, Report No. 110874, August 2013, page 12.
3. "Evaluation of Glass Bottles after Multiple Loops with Regard to their Appearance and Stability", VLB, August 28, 2014, page 7.

Energy and Environmental

OXYGEN ENHANCED NOX REDUCTION (OENR) TECHNOLOGY FOR GLASS FURNACES

Pedel J.[1], Kobayashi H.[2], de Diego Rincón J.[3], Iyoha U.[1], Evenson E.[1], Cnossen G.[1], Zucca P.[4]

[1]Praxair, Inc., 175 East Park Drive, Tonawanda, NY 14150, USA
[2]Praxair, Inc., 39 Old Ridgebury Road, Danbury, CT 06810, USA
[3]Praxair Euroholding S.L., Orense 11 9ª, Madrid 28020, Spain
[4]SIAD S.p.A., 92 via S. Bernardino, 24126 Bergamo, Italy

ABSTRACT

Container and flat glass manufactures in the USA and EU are required to comply with new lower NOx emissions limits as regulation becomes stricter. With modifications of the primary air combustion system such as the optimization of natural gas injection, most furnaces have been able to reduce NOx emissions and comply with current regulations. However, greater reduction (500 to 800 mg/Nm^3 at 8% O_2 dry) was shown to be difficult for many furnaces by the modifications of the fuel injectors alone. Combustion staging to reduce the stoichiometric ratio of the primary air flame by secondary oxidant injection at a strategic location of the furnace has shown significant reduction in NOx emissions in many furnaces. Praxair has developed the Oxygen Enhanced NOx Reduction (OENR) technology to further reduce NOx emissions and achieve NOx levels below 500 mg/Nm^3. A small amount of oxygen is injected near the exhaust ports within the furnace to burn out the remaining CO after optimizing the primary air flame. This paper presents a CFD study of the OENR technology and application examples to cross-fired and end-port furnaces. NOx emissions as low as 380 mg/Nm^3 at 8% O_2 dry flue gas conditions were achieved. Overall, OENR proved to be a cost effective technology, which significantly reduces NOx emissions while maintaining good glass quality and reducing fuel consumption.

INTRODUCTION

Glass making is a very energy intensive process due to the high temperatures required to melt the raw materials. Combustion-generated emissions of nitrogen oxides (NOx) from natural gas (or other fossil fuel) fired glass-melting furnaces are a growing concern as they cause photo-chemical smog, acid rain and depletion of the ozone layer. To reduce fuel consumption and achieve greater thermal efficiency, air is typically pre-heated at temperatures greater than 1300 K through regenerators. This results in peak flame temperatures around 2200 K. The formation of thermal NO is very significant above 1800 K and as a result, the NOx levels from the glass furnaces can typically range from 1000 to 3000 ppm.

Glass manufacturers in the USA and in Europe are pressured by regulators to reduce their NOx emissions. In the USA, the Environmental Protection Agency (EPA) is enforcing more stringent NOx control requirements. The EPA reviewed past furnace rebuild records and adopted a more stringent interpretation of furnace rebuild "triggers". Modifications and upgrades to existing furnace that could be considered as "routine maintenance" in the past are now examined as possible improvements that can extend the life of the plant or its production, and can increase the facility emissions beyond its original Prevention of Significant Deterioration (PSD) permit limit, thus requiring a New Source Review (NSR) permit. Any new furnace construction or major modification to an existing furnace requires a PSD permit from the EPA or the local authorities. To obtain a PSD permit, companies must, among other things, install the Best Available Control Technology (BACT). Currently, the most stringent BACT requirements for NOx are 1.5 lb/ton for container and 3.2 lb/ton for flat glass. The EPA however continues to tighten its standards for BACT. The lowest achievable emission rate technology and technology transfer from other industries are considered to define the new BACT, often regardless of cost effectiveness. Future BACT definitions are therefore expected soon, essentially requiring continuous emission monitoring and add-on technologies.

In Europe, the Integrated Pollution and Prevention Control (IPPC) bureau passed a new Industrial Emission Directive (IED) in 2010 to reduce NOx emissions and implement the Best Available Techniques (BAT) in glass furnaces [3]. The Best Available Techniques Associated Emission Levels (BAT-AELs) essentially requires that most air-fired container and flat glass furnaces achieve NOx levels below 800 mg/Nm^3 (see Tables 1 and 2). The directive is effective since April 2012; all new furnaces must immediately comply and old furnaces will have to comply by 2016/2017, even though regional authorities can negotiate and implement limits on a plant specific basis.

Most air-fired furnaces in the USA and in Europe have already taken some primary measures to reduce their NOx emissions such as excess air reduction, low NOx burners, electric boosting, air leakage sealing or improved furnace design, and are able to achieve NOx emission within the 800 – 1000 mg/Nm^3 range (2.4 – 3.0 lb/ton for container glass). However, greater NOx reduction may prove difficult or costly to achieve for some furnaces by common primary measures. The Oxygen Enhanced NOx Reduction (OENR) technique is a low-cost solution developed by Praxair in collaboration with SIAD to achieve an additional 15 – 40% NOx reduction in regenerative container and flat glass furnaces. This study presents, in a first part, a brief overview of the OENR technical approach, its potential and benefits, and then the results obtained in commercial applications for both cross-fired and end-port fired glass furnaces. Computational Fluid Dynamics (CFD) was used to optimize OENR implementation and gain in-depth understanding of the results.

Table 1. BAT-AELs for NOx emissions from the melting furnace in the container glass sector [1]

Parameter	BAT	BAT-AEL	
		mg/Nm^3	kg/tonne melted glass [1]
NO$_x$ expressed as NO$_2$	Combustion modifications, special furnace designs [2] [3]	500 – 800	0,75 – 1,2
	Electric melting	< 100	< 0,3
	Oxy-fuel melting [4]	Not applicable	< 0,5 – 0,8
	Secondary techniques	< 500	< 0,75

[1] The conversion factor reported in Table 2 for general cases (1,5 × 10⁻³) has been applied, with the exception of electric melting (specific cases: 3 × 10⁻³).
[2] The lower value refers to the use of special furnace designs, where applicable.
[3] These values should be reconsidered in the occasion of a normal or complete rebuild of the melting furnace.
[4] The achievable levels depend on the quality of the natural gas and oxygen available (nitrogen content).

Table 2. BAT-AELs for NOx emissions from the melting furnace in the flat glass sector [1]

Parameter	BAT	BAT-AEL ([1])	
		mg/Nm³	kg/tonne melted glass ([2])
NO$_X$ expressed as NO$_2$	Combustion modifications, Fenix process ([3])	700 – 800	1,75 – 2,0
	Oxy-fuel melting ([4])	Not applicable	< 1,25 – 2,0
	Secondary techniques ([5])	400 – 700	1,0 – 1,75

([1]) Higher emission levels are expected when nitrates are used occasionally for the production of special glasses.
([2]) The conversion factor reported in Table 2 (2,5 × 10⁻³) has been applied.
([3]) The lower levels of the range are associated with the application of the Fenix process.
([4]) The achievable levels depend on the quality of the natural gas and oxygen available (nitrogen content).
([5]) The higher levels of the range are associated with existing plants until a normal or complete rebuild of the melting furnace. The lower levels are associated with newer/retrofitted plants.

OENR PRINCIPLE: STAGED COMBUSTION TO REDUCE NOX EMISSIONS

The Oxygen Enhanced NOx Reduction (OENR) Technology is based on a combustion staging approach and consists in first, reducing the preheated air flow (primary air) while maintaining a similar fuel input to reduce the excess oxygen in the primary combustion zone and then, to inject pure oxygen in a secondary stage to burn any carbon monoxide remaining in the gas stream before it enters the flue port (Figure 1). The effect of reducing the stoichiometric ratio (air/fuel ratio) on NOx formation in glass furnaces has been studied by the Gas Research Institute (GRI) [1, 2]. An almost linear trend has been observed experimentally between the primary stoichiometric ratio and the amount of NOx produced and shows that a significant NOx reduction can be achieved by reducing the amount of excess air (Figure 2). Under normal conditions, glass furnaces operate at a stoichiometric ratio (SR) around 1.1 to maintain an excess oxygen concentration in the flue gas, typically 1.5%. Reducing the stoichiometric ratio by 9% by reducing the air flow would in theory reduce the NOx formation by ~30% according the experimental trend (Figure 2).

However, as the primary stoichiometric ratio is reduced, CO emissions or reducing conditions on the glass surface become the limiting factors for NOx reduction. As shown in Figure 3, CO levels in the flue gas increase dramatically with lower excess air and can reach concentrations greater than 2000 ppm when the O$_2$ concentration in the flue gas drops below 1%. A large amount of unburnt carbon monoxide the flue gas is undesirable because it slightly reduces the energy efficiency of the furnace but, most importantly, because it increases the corrosion of the checkers and can create high temperatures in the regenerator if the unburnt CO reacts with oxygen.

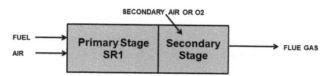

Figure 1. OENR flame staging

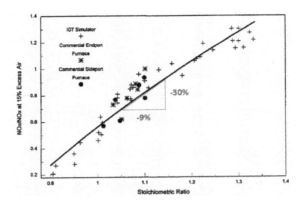

Figure 2. Effect of first-stage stoichiometric ratio on NOx production [1, 2]

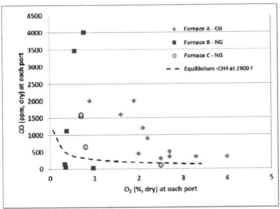

Figure 3. CO emissions vs. excess oxygen

These two effects can significantly reduce the checker life. It is therefore necessary to re-burn the carbon monoxide in an oxygen-rich secondary reaction zone without increasing NOx, which requires good mixing and low temperatures. The other limitation to reducing the stoichiometric ratio is the possible consequences of lower oxygen and higher hydrocarbon concentrations above the glass on the colour and quality of the glass. The presence of unburnt hydrocarbons over the glass and batch surface can change the redox state and the colour of the glass. It can also increase the volatilization of sulphate fining agents and reduce the fining process.

Combustion staging by reducing the stoichiometric ratio of the primary air-gas flame and introducing a secondary oxidant stream at various locations in the furnace has shown to successfully reduce NOx emissions. Four secondary oxidant options work equally well in theory:
- Cold air staging:
  - Simple, but the large air volume disturbs the main flame.
  - Fuel consumption increases since hot air is replaced by cold air.

o   Air preheat temperature increases due to reduced air flow to regenerators.
- Oxygen enriched air staging:
     o   Provide additional flexibility for secondary oxidant penetration and mixing.
     o   More complex equipment, requiring of both air and O_2.
     o   Possible increase in fuel consumption
- Hot air:
     o   Complex to install piping and maintenance concern
     o   Large air volume disturbs the main flame
- Pure oxygen staging ("OENR")
     o   Simple equipment set up and easy to adjust mixing to burn CO on-site.
     o   Reduction in fuel consumption, but additional O_2 cost

Praxair's OENR technology uses a pure oxygen stream to attain lower NOx emissions without adversely impacting the energy efficiency of the furnace or disturbing the air-fuel flame, as is sometimes observed with cold air staging or oxygen-enriched air staging.

OENR MODELLING

Several phenomena need to be taken into account to evaluate OENR performance. First, it is important to understand how reducing the air flow rate and introducing a cold flow of oxygen will affect the regenerators' performance and the air preheat temperature. A small increase in air preheat temperature can indeed have a large impact on NOx formation (Figure 4) [2]. As OENR reduces both the amount of air to preheat and the flue gas volume, NOx reduction could be limited if the air preheat temperature increases. Reducing the primary air/fuel ratio could then lead to an improvement in furnace efficiency and decrease the fuel consumption. GRI study has shown that there is an optimal value of the primary SR around 1.06 where is heat transferred to the glass is maximal (Figure 5) [2]. Therefore, reducing the primary SR from a typical value of 1.12 to values between 1.03 and 1.1 could reduce the fuel consumption and offset the cost of oxygen needed for OENR. Finally, Rue & Abbasi [4] have studied the effect of oxygen concentration in the secondary oxidant and shown that higher NOx reduction can be achieved with higher oxygen levels (Figure 6). Increasing the oxygen concentration in the enriched air from 35% to 50% while maintaining the same primary and overall stoichiometric ratios lead to an additional 10% NOx reduction. A pure oxygen secondary oxidant has thus the potential achieve greater NOx reduction than air staging or enriched air staging.

To assess OENR potential and evaluate the importance of those three effects, Praxair conducted a CFD study using Glass Service© software of a generic end-port furnace with the following characteristics:
     o   300 ton per day (tpd) container flint glass
     o   50% cullet ratio
     o   850 kW electric boost
     o   4 under-port burners

The regenerators were included in the model to account for potential changes in air preheat temperature. The checkers were modelled as a porous media subject to radiative and convective heat transfer. A baseline case with a stoichiometric ratio of 1.12 (12% excess air) was compared to an OENR case with a primary SR reduced to 1.06 and an overall SR kept at 1.12. For both case, the glass throat temperature was controlled and kept constant to 1400 C by a PID controller which adjusted the air and fuel flow rates, so that the reduction in fuel consumption can be estimated. The oxygen injections for the OENR case are located on the opposite side of the flame near the exhaust port.

Figures 7 and 8 show a comparison of the gas temperature profile between the baseline case and OENR (top view and front view). The oxygen jets quickly mix with the surrounding gas and do not create a secondary flame, which would cause high temperatures and generate additional NOx. The main flame is barely disturbed by the oxygen injections; only the flame from the fourth burner (near the centre) is slightly lifted.

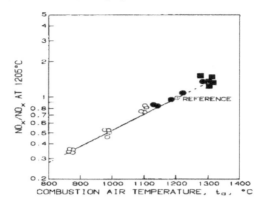

Figure 4. Effect of air preheat temperature on NOx emissions [2]

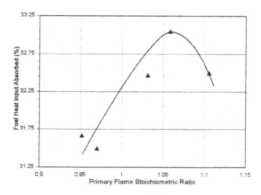

Figure 5. Effect of primary stoichiometric ratio on heat transfer to the glass [2]

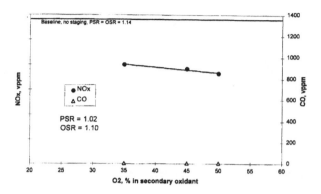

Figure 6. Effect of oxygen concentration in secondary oxidant on NOx [4]

The CFD results are summarized in Table 3. The flue gas volume is reduced by approximately 4% with OENR. The air preheat temperature is mostly unchanged as the lower amount of air to preheat is compensated by a lower flue gas volume and temperature. OENR decreased the fuel consumption by 1% compared to the baseline, which confirms that the fuel consumption won't increase with OENR, as opposed to what has been observed with cold air staging. The NOx were predicted using the Zeldovich mechanism for thermal NOx. The model predicts that with a 6% staging (primary SR reduced by 6%), OENR will reduce the NOx emissions by 25%.

Table 3. Comparison of CFD results between baseline case and OENR

	Base Case	OENR
Air flow rate (Nm3/h)	13950	13250
Natural gas flow rate (Nm3/h)	1340	1320
Oxygen flow rate (Nm3/h)	-	150
Primary SR	1.12	1.06
Pull rate (tdp)	300	300
Throat temperature (°C)	1400	1400
Air preheat temperature (°C)	1152	1148
Exhaust port gas temperature (°C)	1418	1403
O2 concentration in flue gas (%, dry)	1.8	2.2
Flue gas flow (Nm^3/h)	15700	15140
Fuel consumption (%)	100	99
NOx (mg/Nm^3, dry at 8% O$_2$)	875	660
NOx reduction (%)	-	25%

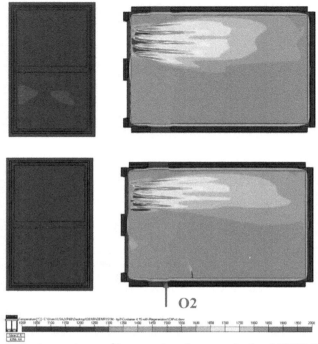

Figure 7: Gas temperature profile comparison: base case (top) and OENR (bottom)

CROSS-FIRED FLOAT GLASS FURNACE APPLICATION

OENR was first tested by Praxair in 1997-1998 as part of a NOx reduction program on a seven-port cross-fired float glass furnace. Float glass typically requires an oxidizing atmosphere above the glass in the combustion chamber. The corresponding excess air usually results in higher NOx levels than for container glass. The furnace had a pull rate of 600 tpd and was operating with oil as a fuel and a 21% cullet ratio. The NOx levels were originally high (~1200 mg/Nm^3) and the furnace was optimized before the OENR test by adjusting the air flow rate at each port and by improving the fuel injectors (oil guns). The new baseline for NOx (1998) was then brought down to ~510 mg/Nm^3.

The OENR tests were done with an under-port oxygen lancing configuration (see Figure 9). The oxygen lances were placed under the exhaust port on the opposite side of the flame with an upward angle.

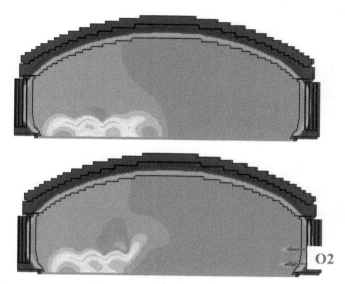

Figure 8. Gas temperature profile in the O_2 injection cross-section (base case on top, OENR on the bottom)

Ports 4 and 5 were identified as the majors NOx contributors because of their high NOx levels and high flue gas volumes. In a first step, it was decided to implement OENR only on ports 4 and 5. The furnace was operating with high CO levels (500 – 3000 ppm) in the exhaust ports and the objective of this first OENR test was to reduce the combustion air while controlling the CO levels in the ports. Preliminary tests were done first on a single port to optimize the flowrate, angle and velocity of the oxygen jet. Proper velocity and momentum proved critical to avoid disturbing the flame and ensure good mixing with the flue gas stream.

For this first test, the excess air in ports 4 and 5 were gradually reduced and the oxygen flow rate increased to 60 Nm^3/h for port 4 and to 80 Nm3/h for port 5. The results are presented in Table 4. The CO levels in port 4 were reduced from 350 ppm to 150 ppm and from 1000 ppm to 300 ppm in port 5. The overall NOx emissions were reduced from 510 ppm down to 490 ppm only.

Primary combustion zone   **Underport OENR**

Figure 9. OENR configuration for cross-fired furnaces

OENR was then extended to ports 3, 4, 5 and 6 for the second test. The final oxygen flow rates were respectively 75, 65, 110 and 60 Nm3/h for ports 3, 4, 5 and 6. Results for NOx and CO levels are presented in Table 4 and Figure 10. The average CO concentration in the stack was 26 ppm and the NOx concentration was reduced by 20% from 480 mg/Nm3 to 380 mg/Nm. The glass quality and colour were not affected by the change in air stoichiometry.

Table 4. Comparison of NOx and CO levels with OENR to the baseline

		NOx in stack (mg/Nm3 at 8% O2)	CO in stack (ppm)	CO in ports (ppm)
Test 1	Baseline	510	< 100	500 - 3000
	OENR (2 ports)	490	< 100	100 -300
Test 2	Baseline	480	< 100	
	OENR (4 ports)	**380**	< 100	

Overall, these two tests showed that OENR has the capability to burn out excess CO levels entering the regenerators and to reduce NOx emissions from an initially low level, without affecting the glass quality, even for a white glass requiring oxidizing conditions. An average NOx concentration of 380 mg/Nm^3 (dry, 8% O$_2$), corresponding to a 20% NOx reduction, was achieved with a total oxygen consumption of 310 Nm^3/h for a 600 tpd float furnace.

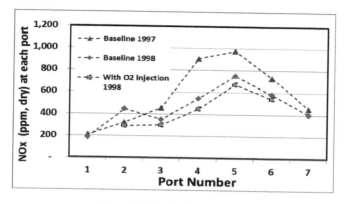

Figure 10. NOx levels at each port

END-PORT FURNACE APPLICATION

More recently, in 2013, Praxair had the opportunity to collaborate with SIAD to implement OENR in an end-port regenerative glass furnace. The furnaces produces high quality white soda-lime glasses for perfume bottles with a pull rate varying from 50 to 80 tpd. The furnace uses natural gas and no electric boosting. The customer is facing new NOx limits that will come into effect in 2016 under the European directive. The OENR test was performed to reduce NOx emissions below the future limits.

To perform the test, oxygen lances were placed at different locations in the furnace. The customer did not want to drill holes in the sidewalls so the lances had to be inserted through the peepholes. Four oxygen injection locations were possible (Figure 11). As in any regenerative furnace, the flame alternates sides between left firing and right firing conditions. It is thus possible to inject the oxygen on the firing side or in the opposite side. The customer was interested in evaluating both, thus the following configurations for oxygen injection were considered:

- o   2 & 3 on the firing side
- o   2 & 3 on the exhaust side

Oxygen flow rates varied from 50 Nm³/h to 130 Nm³/h. The oxygen concentration in the exhaust port was maintained constant to ~1.5% (dry) by adjusting the primary air/fuel ratio. Despite the CFD results showing that greater NOx reduction could be achieved with locations 1 & 4, the customer was concerned about creating reducing conditions above the glass surface and this configuration could not be tested. The tests were conducted during normal operation of the furnace without affecting the glass production or quality. Regular gas sampling was done in the ports after reaching stable conditions. Stack emissions and optimal measurements of the crown temperature were also collected.

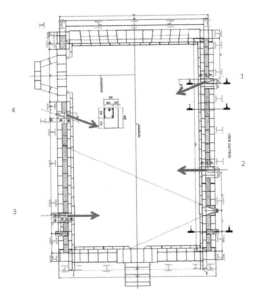

Figure 11. Various oxygen injection locations in the end-port furnace

The results for NOx emissions were plotted against the natural gas consumption for various conditions as the glass pull rate varied significantly (Figure 12). The injection of 50 Nm3/h of oxygen on the firing side lead to a negligible reduction in NOx, and increasing the flow rate to 130 Nm^3/h actually increased the NOx emissions above the baseline level. The best results were achieved with the injection of 50 Nm^3/h of oxygen on the exhaust side. This configuration reduced the NOx concentration by 150 mg/Nm3 on average compared to the baseline, which translates into a NOx reduction of 15 to 18%. This performance was fairly constant over the range of natural gas input. Higher oxygen flow rates of 100 or 130 Nm3/h achieved lower NOx reductions. Even though a deeper staging should lead to lower NOx levels, it is probable that the higher oxygen flow rates injected at this location caused the oxygen jet to mix with the flame and thus limited the amount of NOx reduction.

Figure 13 presents the concentrations of carbon monoxide measured in the port for OENR with injection of oxygen on the exhaust side. In general, an increase in CO concentration is observed with OENR. The CO concentration ranged from 900 to 1100 mg/Nm3 on average for OENR with 50 Nm3/h of oxygen, which was acceptable for the customer. The higher flows of oxygen provided a better control of the CO levels, even though the NOx performance was not as satisfactory.

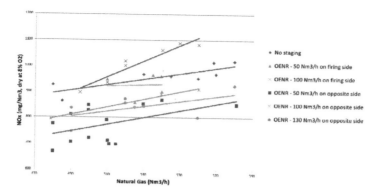

Figure 12. NOx emissions in stack vs. natural gas consumption for different OENR configurations

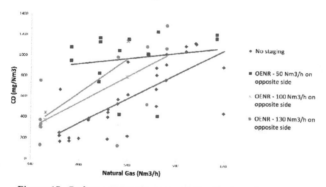

Figure 13. Carbon monoxide concentration in the exhaust port

Temperature measurements of the crown temperature were also recorded. Figure 14 and Figure 15 compare the crown temperature profile between normal operations and OENR conditions. Overall, OENR did not create any hot spot on the crown and kept the temperature profile similar to the no staging conditions. Furnace operators mentioned that the heat distribution inside the furnace might actually be improved with OENR. No issue in glass quality or colour aroused during the whole test. It was not possible to accurately measure any change in fuel efficiency as the air and fuel flow rates were not measured accurately and the pull rate frequently changed, but no adverse effect was observed as the air preheat temperature and the flue gas temperature remained similar.

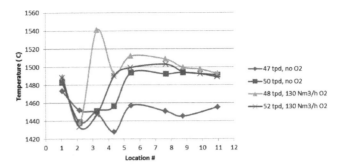

Figure 14: Comparison of crown temperature profile between no staging conditions and OENR with 130 Nm^3/h O^2 (exhaust side)

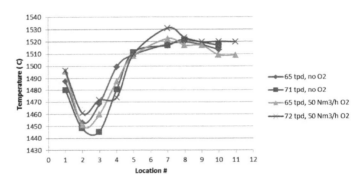

Figure 15: Comparison of crown temperature profile between no staging conditions and OENR with 50 Nm^3/h O$_2$ (exhaust side)

CONCLUSION

In a context of stricter NOx regulations around the world, Praxair has developed the Oxygen Enhanced NOx Reduction (OENR) technology in collaboration with SIAD to reduce NOx emissions in regenerative glass furnaces. OENR is a combustion staging technique, which is based on reducing the primary air/fuel ratio and burning out the remaining CO with a secondary injection of oxygen. The technology has been validated both through computer simulations and commercial scale tests. The reduction of combustion air and the low amount of oxygen injected into the furnace reduce the flue gas volume without disturbing the flame, and could in theory improve the furnace efficiency. OENR has been implemented in cross-fired float glass and end-port white soda-lime glass furnaces. The technology has proven successful in reducing NOx emissions while maintaining glass production and quality. Its implementation is simple for the furnace operators as it does not change significantly the heat distribution in the furnace. Overall, due to the simplicity of its installation and the small amount of oxygen needed, OENR is a low cost solution to achieve NOx reduction from 15% up to 40%.

REFERENCES

[1] H. Abbasi, M. Khinkis and D. Fleming, "Evaluation of NOx emissions on pilot-scale furnace," in *44th Annual Conference on Glass Problems*, 1983.

[2] H. Abbasi, D. Fleming and H. A. Abbasi, "Development of NOx control Methods for Glass Melting Furnaces," Institute of Glass Technology, 1987.

[3] European Commission, "Best Available Techniques Reference Document for the Manufacture of Glass," 2010.

[4] D. Rue and H. Abbasi, "Demonstration of oxygen-enriched air-staging at Owens-Brockway glass containers," Institute of Gas Technology, 1997.

U.S. AIR REGULATIONS INVOLVING GLASS MANUFACTURING

Steven B. Smith
EHS & Regulatory Affairs Consultant
Muncie, IN 47302, USA

ABSTRACT

Basics of environmental rulemaking is reviewed as well as the role of the states. An overview Part 70, Title V permitting is provided, what various Title V permits exist today, why permit variations occur and when a state should be notified about work on a permitted process. Greenhouse gas requirement are in place today at the Federal level and at the state level one state has set limits and requirements for controls but various agreements foretell expansion of such rulemaking to other jurisdictions.

INTRODUCTION

Glass manufacturing uses large amounts of raw materials and fuels to melt those materials in the production process. Glass is a commodity product for the most part, which means that large quantities of finished products must be produced to manage the business and to satisfy the customer demand. Because of all this, environmental rules or laws that a state manages may be triggered and air permitting may be required.

BASICS OF ENVIRONMENTAL RULEMAKING

A property owner on a lake may own property beyond the shoreline, for example out to the middle of the lake. That owner can erect fences on his property and restrict access to the property. However, the owner cannot stop a person in a kayak or speed boat from crossing over that property on the water, so long as the land is not physically touched. That is because the water is owned by the state. The same principle works for environmental air rules and how they work. If a business purchases land and builds a glass plant, the business can build a fence and restrict access to the property. However, the business cannot restrict drones or aircraft from flying overhead because the state owns the air. Control of the air above the property remains with the state or Federal government. This ownership issue is the basis for most environmental rules. Because the state owns the air, it may set thresholds for specific materials that a factory might want to emit and if that factory is over that threshold amount, the state can have say in how much is emitted and the state can also control parts the process of manufacturing itself. Almost all the Rules that a facility may have to comply with today are state Rules. The U.S. EPA creates and develops Rules that each state must enforce within that state. Consequently there are many variations to any one Rule that the EPA has set. Because of the large amount variability among all the state Rules, this discussion will focus primarily on the Federal Rules.

CLEAN AIR ACT, PART 70 – TITLE V AIR PERMITS [1]

Title V air permits are:

- Designed to regulate Criteria Pollutants, see chart below.
- Issued by the state but,

- Are legally enforceable by both the state and the Federal governments.
- Written to include any required manufacturing controls and written to include any control equipment.
- Different from state to state and can be straightforward and detailed to highly detailed and very complex for very similar manufactures.

The U.S. EPA has set a list of Criteria Pollutants that is based on a facilities 'potential to emit' or PTE, as shown in Table 1. Various states and areas may have different nomenclature for this. It is in essence, the design output of a permitted source times the total hours in a year, as though that source ran at 100% production for a full year. Emissions are then calculated based on that production rate. It takes but one of these pollutants over the threshold to trigger this level of permitting. Note that this is different than the actual emissions that a facility might emit. It should also be noted that states may impose more restrictive threshold levels for a number of reasons and many have done that.

Table 1. Permitting thresholds of different pollutants

Part 70 Permitting thresholds, based on potential to emit, in Tons per Year (TPY)

Pollutant	Threshold
Volatile Organic Compounds (VOCs)	100 TPY
Carbon Monoxide (CO)	100 TPY
Nitrogen Oxides (NO_x)	100 TPY
Sulfur Dioxide (SO_2)	100 TPY
Fine Particulate Matter (PM_{10})	100 TPY
Very Fine Particulate Matter ($PM_{2.5}$)	100 TPY
Combined Hazardous Air Pollutants (HAPs)	25 TPY
Carbon Dioxide Equivalents (CO_2e)*	100,000 TPY
Individual HAPs	10 TPY each
Lead	10 TPY

*As of 7/1/2011

There are essentially two types of Title V air permits [2].

- Minor source.
  Referred to as 'non-Title V' or 'Synthetic Minor Title V' on depending on whether or not the sources took voluntary restrictions on the quantity of criteria pollutant emissions. For example:
    1. A source has PTE emissions of NOx of 90 tons per year and the Part 70 Permitting threshold is 100 tons per year. This source would qualify for a non-Title V permit.
    2. A source has PTE emissions of NOx of 105 tons per year but they always shut down for the Christmas Holiday. That source can take restrictions to shut down for the Holiday to reduce their PTE below the Part 70 Permitting

threshold. This source would then qualify for a Synthetic Minor Title V permit. It should be pointed out that once this restriction is in the permit, the source would be required to comply and even with changing business conditions could not then run during this time unless the permit was modified.

- Major Source.
  This level of permit is generally referred to as a 'Major Source' and/or 'Title V Facility' depending on what stage they are in the permitting, installation and operation process. The facility wide potential to emit is above at least one of the threshold levels and triggers a major source permitting requirement. The actual type of permit issued to these sources depends on when a given operation at the facility was installed, the state in which the operation is located and whether or not other restrictions may come into play.

During the early 1990's as a part of the initial implementation of Title V permitting the state regulators openly discussed these permits as being intentionally difficult to attain and to manage. The objective of making these difficult was that industries would seek alternate ways to manufacture with fewer emissions. Today, it is apparent that this goal was achieved, that these permits are difficult to very difficult to attain and to manage. In addition to that, each state has their own approach, methodology and paperwork requirements. Interpretation of specific requirements can vary by region, by state, or in areas within a state. Most manufacturing Title V permits, including those for glass manufacturing, can be found on the respective state web sites.

One specific aspect or Title V permitting needs also be mentioned. When is it required to notify the state and when are permit modifications required? According to the U.S. EPA, state notifications are required anytime:

- Any work is done beyond maintenance on a permitted source
- Any changes or modifications to the process – especially if a production bottleneck is removed.
- Modifications or physical change which increases the amount of any air pollutant emitted or which results in the emission of any air pollutant not previously emitted.

Whether or not a permit or a permit modification is necessary is the state's decision depending on the specific case variables. A case that requires a permit modification in one state may not require a permit modification in a different state. The state must be notified in order for the state to evoke the right to make the permit requirement or modification decisions.

U.S. EPA MENU OF CONTROL MEASURES [3]

The U.S. EPA has published its Menu of Control Measures. This document lists the various criteria pollutants and the control measures currently technically feasible or acceptable for that pollutants control. It also lists these by individual manufacturing industries including glass manufacturing. Also included in the document is an estimated cost per ton of pollutant removed although the values listed in the document appear to be understated.

It would behoove any glass manufacturer that is preparing to have discussions about control equipment with a regulator to attain this document and be prepared to discuss or to demonstrate a more realistic cost benefit analysis for the various types of control equipment. The U.S. EPA Menu of Control Measures document can be found at: www.epa.gov/air/pdfs/menuofcontrolmeasures.pdf

NATIONAL EMISSIONS STANDARDS, HAZARDOUS AIR POLLUTANTS, NESHAPS [4]

The U.S. EPA has listed 190 materials that are listed as Hazardous Air Pollutants or HAPs. These are materials that are purposely introduced into production as raw materials and may be released as air emissions. A glass manufacturing facility that has any of these materials in its production process may have additional requirements for these materials in its Title V permit. For glass manufacturing, additional details and requirements can be found in Subpart SSSSS [5]. Some common HAPs used in glass manufacturing are:

- Arsenic compounds
- Antimony compounds
- Cadmium compounds
- Chromium compounds
- Cobalt compounds
- Lead
- Manganese
- Nickle

NATIONAL AMBIENT AIR QUALITY STANDARDS, NAAQS [6]

The U.S. EPA has also published a listing of ambient air requirements that states must attain and manage, as shown in Table 2. States are required to sample ambient air in suspect areas and report this information to the EPA. Areas where the standards are not met must be addressed and the state must write a State Implementation Plan or SIP, outlining how and when the state will reduce the levels to attain the limits. Areas not meeting the ambient air requirements are called nonattainment areas. As a part of the SIP the state generally lowers the criteria pollutant threshold for that area, usually done by county or parts of a counties. Once it is submitted to the EPA for approval the EPA may expand the nonattainment area to adjacent counties of up to 50 kilometers, or about 31 miles to allow for a broader area to help control the pollutant. This expansion can also cross state lines into neighboring states. All of this has the effect of making Title V permitting much more difficult for any sources emitting that particular pollutant. It is also possible that the state could ask a facility to install control equipment to remove tons of the specific pollutant involved in the nonattainment issue. This can occur to a facility even though there has not been a change to the process, to emissions or to equipment. Figure 1 shows a map of the U.S. showing where those area are that are nonattainment or have been nonattainment are now having to work to maintain the attainment level [7].

Table 2. List of ambient air requirements

Pollutant [final rule cite]		Primary/ Secondary	Averaging Time	Level	Form
Carbon Monoxide [76 FR 54294, Aug 31, 2011]		primary	8-hour	9 ppm	Not to be exceeded **more** than once per year
			1-hour	35 ppm	
Lead [73 FR 66964, Nov 12, 2008]		primary and secondary	Rolling 3 month average	0.15 μg/m³ [1]	Not to be exceeded
Nitrogen Dioxide [75 FR 6474, Feb 9, 2010] [61 FR 52852, Oct 8, 1996]		primary	1-hour	100 ppb	98th percentile, averaged over 3 years
		primary and secondary	Annual	53 ppb [2]	Annual Mean
Ozone [73 FR 16436, Mar 27, 2008]		primary and secondary	8-hour	0.075 ppm [3]	Annual fourth-highest daily maximum 8-hr concentration, averaged over 3 years
Particle Pollution Dec 14, 2012	PM₂.₅	primary	Annual	12 μg/m³	annual mean, averaged over 3 years
		secondary	Annual	15 μg/m³	annual mean, averaged over 3 years
		primary and secondary	24-hour	35 μg/m³	98th percentile, averaged over 3 years
	PM₁₀	primary and secondary	24-hour	150 μg/m³	Not to be exceeded more than once per year on average over 3 years
Sulfur Dioxide [75 FR 35520, Jun 22, 2010] [38 FR 25678, Sept 14, 1973]		primary	1-hour	75 ppb [4]	99th percentile of 1-hour daily maximum concentrations, averaged over 3 years
		secondary	3-hour	0.5 ppm	Not to be exceeded more than once per year

Counties with Nonattainment Areas or Maintenance Areas
for Clean Air Act's National Ambient Air Quality Standards (NAAQS) *

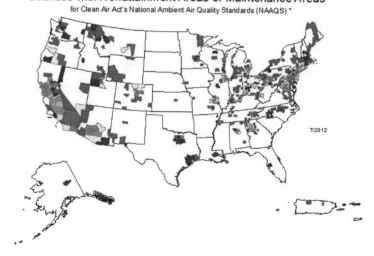

7/2012

Figure 1. A map of the U.S. showing areas of nonattainment or nonattainment in work

GREENHOUSE GAS REGULATIONS

No discussion about U.S. air regulations would be complete without covering greenhouse gas regulations. This is certainly an emerging area of rulemaking. On 11/12/14 The US announced an agreement with China to cut GHG emissions 26-28% of 2005 levels by 2025 [8]. The effects of this agreement have yet to be seen. Today, the U.S. EPA requires that greenhouse gas (GHG) be reported by any facility emitting 25,000 metric tonnes, (mtonnes), in an electronic system called the Electronic Greenhouse Gas Reporting Tool or e-GGRT. On its cover page for its GHG web site the EPA says [9]:

> *EPA's Greenhouse Gas Reporting Program will help us better understand where greenhouse gas emissions are coming from and will improve our ability to make informed policy, business, and regulatory decisions.*

This statement appears to outline that the EPA intends to use the data collection information to make decisions about rulemaking going forward. Today maps of US emissions, glass industry totals and individual facility emissions information are all available on the EPA web site.

At the state level, one state has passed GHG legislation and has an active program in place. California passed AB32, the Global Warming Solutions Act [10]. This law has in place today requirements that:

- Facilities with 25,000 mtonnes are required to participate in the program.
- Requires mandatory GHG reporting
- Contains a validation process for reporting of GHG emissions
- Has an operational Cap & Trade program.
- Has set statewide eroding limits for overall state wide emissions
- Sets benchmark levels for individual industries that includes the statewide eroding emissions

The current Rule runs through 2020. California will need to pass an updated Rule to continue this process beyond 2020 and planning for that and how it might function has now begun. On 5/15/14 the California Air Resources Board, CARB published an Update to the Scoping Plan outlining the intended changes to the 2020 Rule and how it might look through 2050. Figure 2 shows the annual GHG emission as a function years in decades [11]), as included in that Scoping Plan update. It shows that the state intends to develop guidelines to increase the downward slope for GHG emission after 2020. This would result in accelerating the decreases in emissions that California facilities emit, including that of the glass manufacturing businesses.

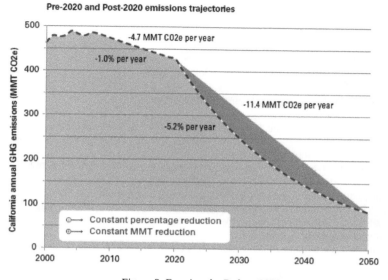

Pre-2020 and Post-2020 emissions trajectories

Figure 2. Framing the Path to 2050

The possibility of the California AB32 GHG program expanding to other states is realistic and that is a goal that the California regulators openly discuss. On 1/1/14 Quebec, Canada formally aligned with California's Cap & Trade program [12]. In July of 2014, Quebec introduced legislation to 'more fully harmonize' their program with California's program.

On 10/28/13 Governors from California, Oregon, Washington and the First Minister from Canada's British Columbia met at the Pacific Coast Action Plan on Climate and Energy [13]. One outcome was that they all signed an agreement to align GHG rulemaking. Oregon has introduced legislation for this but it has yet to come to fruition [14]. Washington State's Governor has been very active with this effort, holding Public Hearings on GHG Rulemaking and has formed a bi-partisan Task Force to process GHG legislation [15]. The final report from this Task Force is due on the Washington Governor's desk on 11/22/14.

REFERENCES

1. Clean Air Act Title V requirements found at:
   http://www.epa.gov/airquality/permits/obtain.html
2. Clean Air Act Title V permitting requirements found at: :
   http://www.epa.gov/airquality/permits/guidance.html
3. EPA Menu of Control Measures found at:
   www.epa.gov/air/pdfs/menuofcontrolmeasures.pdf
4. EPA NESHAP found at:
   http://www.epa.gov/compliance/monitoring/programs/caa/neshaps.html
5. EPA NESHAP Subpart SSSSSS found at: http://www.epa.gov/ttn/atw/area/arearules.html
6. EPA NAAQS found at: http://www.epa.gov/air/criteria.html
7. EPA NAAQS US Map found at: www.EPA.gov/air quality/greenbook/mapnmpoll.hytml
8. US, China GHG agreement found at: http://www.whitehouse.gov/the-press-office/2014/11/11/us-china-joint-announcement-climate-change
9. US EPA Greenhouse Gas Program found at: http://www.epa.gov/ghgreporting
10. California AB32 Global Solutions Warming Act found at:
    http://www.arb.ca.gov/cc/cc.htm
11. California ARB Scoping Plan Update found at:
    http://www.arb.ca.gov/cc/scopingplan/document/updatedscopingplan2013.htm
12. CARB announcement of Quebec aligning with CA Cap & Trade found at:
    http://www.arb.ca.gov/newsrel/newsrelease.php?id=657
13. The Pacific Coast Action Plan on Climate and Energy found at:
    http://www.pacificcoastcollaborative.org/Documents/Pacific%20Coast%20Climate%20Action%20Plan.pdf
14. Oregon GHG Bills found at: https://www.oregonlegislature.gov/
15. Washington GHG Task Force found at:
    http://www.governor.wa.gov/issues/climate/cert.aspx

NEW COMBUSTION TECHNIQUE FOR REDUCING NOx AND CO_2 EMISSIONS FROM A GLASS FURNACE

Pont R S, Global Combustion Systems Ltd, UK
Fricker N, University of South Wales, UK
Alliat I, GDF SUEZ-CRIGEN, France
Agniel Y, O-I Manufacturing France, France
Kaya L, Turkiye Sise Cam, Turkey

ABSTRACT

The European Glass Industry has the problem of reducing both CO_2 and other emissions such as NO_x from its primary regenerative melting furnaces. To address this problem a new combustion concept has been developed by a consortium of GDF-SUEZ, Global Combustion Systems and the University of South Wales with financial assistance from the UK's Carbon Trust. It comprises a novel, patented firing technique (AUXILIARY FIRING) that reduces NOx formation at source on primary regenerative glass melters, while simultaneously reducing fuel consumption and CO_2 emissions. It avoids or reduces the need for post-furnace NOx clean up which can be expensive in capital and increases life-cycle CO_2 emissions.

If confirmed over longer term testing and on cross-fired furnaces, the results of this project will enable the European glass industry to meet upcoming NOx emission limits on their primary glass melters by reducing NOx formation at source, without the need to install large, expensive and energy-hungry post furnace clean up techniques. Elimination of NOx clean up will yield simultaneous reductions in NOx and glass life-cycle CO_2 emissions. This paper outlines the technique, optimisation using modelling methods and the results obtained when applied to industrial scale glass furnaces at O-I Manufacturing France and SISECAM Turkey. The research leading to these results has received funding from the European Union Seventh Framework Programme (FP7/2007-2013) under grant agreement n°296042

INTRODUCTION

The first part of the development of the application to regenerative Glass furnaces was carried out under a UK Carbon Trust funded project the results of which were more related to cross-fired furnaces. The second part funded by an EU FP7 grant involved application to both Cross-fired and end-fired furnaces with Trials on typical industrial glass furnaces. Traditionally regenerative glass furnaces are fired with turbulent diffusion flame burners located in or under the airports. This causes initial rapid mixing of hot combustion air and fuel resulting in high temperatures in the flame envelope generating high concentrations of NOx.

The principle of this development is auxiliary combustion in which a second (Auxiliary) fuel jet is introduced into the furnace in such a way that it burns in the excess oxygen contained in combustion products circulating within the furnace. This produces an auxiliary flame free from the high peak temperatures of the conventional flames from underport or through port burners resulting in lower NOx. The auxiliary flame also has increased radiative properties resulting in a flame with improved heat transfer. At the same time the conventional burners are firing at a lower rate with high excess air resulting in lower NOx generation. The net result is lower NOx and improved heat transfer to the glass. The combination of the conventional and auxiliary burners allows tuning of the furnace for the best compromise between thermal efficiency and NOx emissions by adjusting the proportion of conventional and auxiliary fuel. Auxiliary combustion requires very careful selection of the injection point, direction and velocity of the fuel. In order to study this some form of flow modelling was required.

Auxiliary combustion should not be confused with Staged combustion, which has been applied to Regenerative furnaces in order to reduce NOx and improve efficiency [1]. Staging involves injecting fuel into the path of the combustion air further along the combustion path from the primary burners whereas the process being investigated here requires entrainment of combustion products into the fuel stream. Furthermore Staging proportions are usually small compared Auxiliary Combustion.

MODELLING

In order to understand the flow patterns within the combustion chamber to allow selection of suitable Injection points and make initial evaluation of the likely effect, flow modelling was required. Both computational fluid dynamics (CFD) and acid/alkali physical modelling were used.

Acid/Alkali Modelling proved to be an excellent technique for establishing satisfactory injection points considering any restrictions imposed by the furnace design and construction. Acid/alkali modelling [2] requires a scale model of the furnace and uses a weak acid to represent the combustion air and a weak alkali containing a pH sensitive indicator e.g. methyl blue to represent the gas. When the acid (air) mixes with the alkali (gas) in the furnace neutralisation occurs as the alkali entrains the acid and so the indicator becomes clear thus simulating the flame envelope. Effectively this is the same process as "mixed is burnt" occurring in turbulent diffusion flames. The system is shown diagrammatically in Figure 1 and a typical facility is shown in Figure 2. The technique has proved to provide a very accurate dynamic simulation of the flames in a glass furnace.

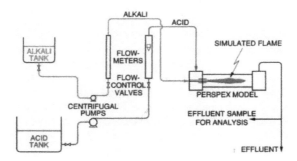

Figure 1. Acid/alkali modelling system

Figure 2. Acid/alkali modelling lab at USW

Figures 3 and 4 show under port combustion in an end-fired and cross-fired furnace. This modelling confirmed it was possible and desirable to combine the conventional firing system with the Auxiliary combustion process, thus providing flexibility and the possibility of optimising the performance of the two systems.

Figure 3. Acid/alkali model of under port flames in an end-fired furnace

CFD was used to obtain quantitative data relating to the performance. Unfortunately full CFD glass tank combustion models take a long time to converge. However, simplified Isothermal CFD proved valuable in determining the flow data for zone modelling.

Figure 4. Acid/Alkali model of under port flames in a Cross-fired furnace

Zone Modelling [3] divides the combustion space into a number of zones. The flow in and out of each zone is determined by isothermal CFD and the boundary conditions are set for the furnace. The basic process of Zone modelling is shown in Figure 5.

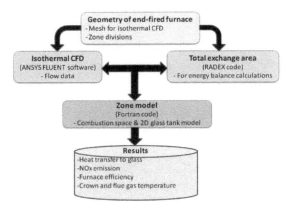

Figure 5. Flow diagram for zone modelling

Zone modelling is used to carry out parametric studies of fixed geometries. Because Zone modelling is very quick a large number of conditions can be investigated in a short time. This technique will be used to predict the performance of applications.

PILOT TESTING

Auxiliary combustion is very different from conventional methods of firing therefore it was essential to test it on pilot scale furnaces. Such tests would allow different firing conditions and parameters to be tested and provide real performance data, which it was hoped would be representative of the performance on a full sized industrial glass furnace. Furthermore such tests would help to convince a glass maker that it would be useful and safe to test on a real glass furnace. Figure 7 shows the internal shape of the 2MW test furnace. The glass surface was simulated by a layer of ceramic fibre over a series of water-cooled tubes, as shown in Figure 8.

Figure 6. GDF SUEZ-CRIGEN 2MW test furnace

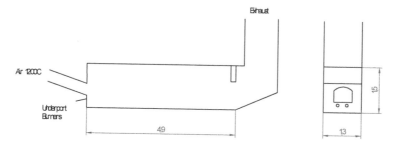

Figure 7. Internal shape of the 2MW test furnace

Figure 8. Hearth of the 2MW test furnace

The non-reversing test furnace used combustion air electrically preheated to 1100C and conventional dual-impulse under port burners fired on natural gas. Water-cooled injectors (called Auxiliary Injectors) were used to allow the gas for auxiliary combustion to be injected in many locations and directions.

Instrumentation was provided to measure structure and gas temperatures and to analyse for O_2, CO, NO, NO_2, and CO_2. The Water flow and temperature through the hearth cooling tubes were measured to allow calculation of the heat flux and distribution to the hearth.

The tests demonstrated certain configurations could result in a very significant reduction in NOx and improvement in heat transfer to the hearth ("glass") Figures 9 and 10 The % AI is the percentage of the total gas entering the furnace via the auxiliary injectors the remainder being fired through the under port burners.

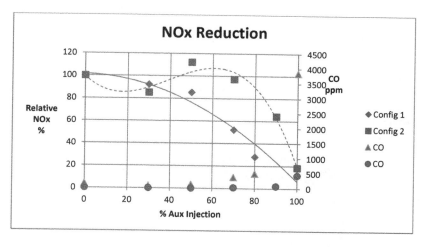

Figure 9. NOx in tests of different auxiliary injection configurations

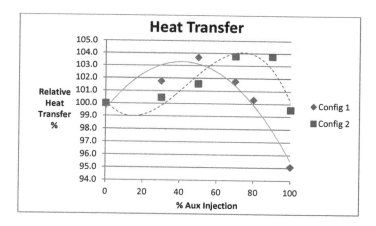

Figure 10. Heat transfer in tests of different auxiliary injection configurations

These results and our other observations may be summarised as follows:
- NOx can be reduced to by 75% with a small increase in CO
- Heat transfer can be increased
- Crown temperature may be reduced or increased
- The above can occur simultaneously in varying degrees

- The Auxiliary Injector configuration is very important in achieving the desired result therefore some method of identifying the optimum configuration is desirable

Because end-fired furnaces have very different and more complex flow patterns compared with cross-fired furnaces, the model studies were repeated for typical end-fired furnaces. A one fifth scale pilot End fired furnace was constructed in the GDF SUEZ-CRIGEN test facility (Figure 11) with a heat input of 500 kW. The design was based on the furnace on which the Industrial tests will be carried out.

Figure 11. New end-fired pilot furnace at the CRIGEN laboratory of GDF SUEZ

The pilot furnace was fitted with dual impulse under port burners, used combustion air at 900-1100°C and had a water-cooled hearth covered with ceramic fibre.

Figure 12. GDF SUEZ end-fired pilot furnace gas flame in conventional combustion mode

Similar instrumentation to that used on the cross-fired pilot furnace was used. Testing the proposed auxiliary combustion injectors suggested by the acid/alkali modelling demonstrated that significant NOx reduction could be achieved as shown in Figure 13.

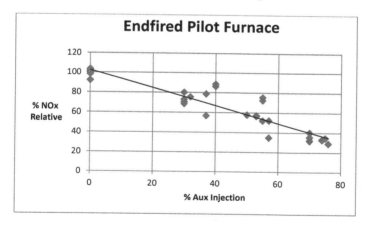

Figure 13. NOx in the GDF SUEZ Pilot furnace exhaust

For this furnace, the base case NOx without auxiliary Injection was 1000 mg/m^3 and CO around 50 ppm. All the results shown in the graph are based on the injector being located in what was considered the optimum position although many other locations and configurations were tested. This location was also a practical position for the Industrial furnace tests.   With Auxiliary Injection CO was always below 200 ppm under all reasonable conditions of Auxiliary Injection. Variations in NOx are largely due to the various Auxiliary Injector configurations.

Heat transfer to the hearth ("glass") varied depending on configuration although there was a general trend towards greater efficiency (see Figure 14) as the auxiliary injection increased as had been observed on the 2MW pilot "cross fired" furnace.

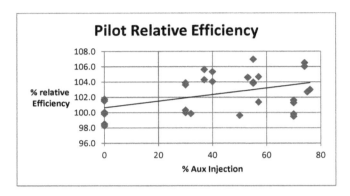

Figure 14. Heat transfer efficiency in pilot furnace tests

INDUSTRIAL TESTING

Funding provided by the EU FP7 programme allowed two commercial glassmakers to join the development thus providing facilities for industrial testing of the process. One glass makers could test on two end-fired container furnaces fitted with dual impulse under port burners and the other could test on an Cross fired float furnace fitted with side of port neck dual impulse gas burners.

All the furnaces were already fitted with chimneystack monitoring systems. To provide additional data to allow the process to be better understood and to confirm and refine mathematical models being developed measurement of O$_2$, CO, CO$_2$ NOx, and gas temperatures were carried out in the regenerator top and bottom on a continuous basis during testing.

INDUSTRIAL TESTING RESULTS – END-FIRED FURNACES

EF1 Furnace - This was a relatively new highly efficient 98m^{2}furnace melting 300 t/day of green container glass. The initial testing of the End fired furnace covered a period of two weeks during which time various Injector parameters could be adjusted as well as adjusting the conventional burners to optimise the process. As in the pilot furnace tests water cooled Auxiliary injectors were used to maximise flexibility for injecting the auxiliary combustion gas.

Glass temperature and quality were the main conditions to be maintained during testing while adjustments were made to the firing system to minimise NOx and keeping CO in the top of the regenerators below 300 ppm. The tests on the End fired furnace confirmed that NOx could be reduced as indicated by the pilot tests see Figure 15.

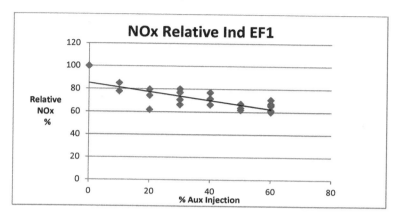

Figure 15. NOx reduction in EF1 test

Longer term testing is required to evaluated effects on the thermal efficiency of the furnace while operating at the regulated NOx level of 500 mg/m^3. The furnace on which these tests will be carried out has a very stable pull rate and is also one of Europe's most efficient container furnaces so it should be possible to detect changes in efficiency of ± 2%.

EF2 Furnace - This is slightly larger End fired furnace than EF1 although geometrically similar, melting clear container glass at up to 450 t/d. This furnace has a high and very variable pull rate and considerably higher NOx levels than EF1. The furnace is under port fired with dual impulse burners. Testing was carried out over ten days. During this time NOx, CO, O$_2$, were continuously measured in regenerator tops and exhaust flues. The exhaust and combustion air temperatures were also measured continuously in the top of the regenerator chambers. Water cooled Auxiliary Injectors were used as in EF1.

During the tests there was considerable variation in pull and gas input, which combined with the adjustment of Auxiliary Injection operating conditions resulted in a high degree of variability in the results. Furthermore the amount of auxiliary gas, which could be applied was limited by the supply system especially at high gas rates.

As a result the data shown in Figure 16 has been selected to show the most stable data at an intermediate and stable gas input. In all the cases shown the CO was below 300 ppm. Again there was a similar trend to that observed on EF1 and the Pilot furnace. Due to the variation in furnace operating conditions it was impossible to evaluate any changes in furnace thermal efficiency.

A further series of tests are proposed for EF2 using the best configurations from the first series of test. Modification to the gas supply system should allow Auxiliary Injection at higher gas rates.

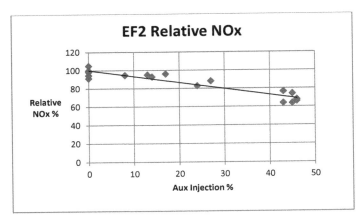

Figure 16. NOx reduction on EF2

Although the three furnaces, Pilot, EF1 and EF2 were of different sizes and operated under different conditions with different Base case NOx levels varying from 600 to 1000 mg/m^3(@8%O$_2$) the Reduction in NOx relative to the amount of auxiliary injection was very similar as shown in Figure 17.

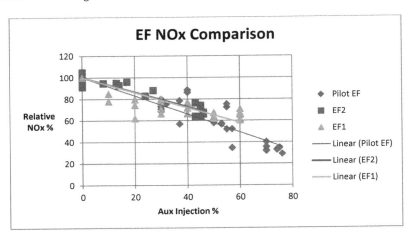

Figure 17. Comparison of NOx results for three furnaces

In all three cases the location and configuration of the Auxiliary Injection was similar indicating that the process can be applied to a wide range of furnace sizes and operating conditions. It should be noted that in the cases of the Industrial furnaces the Auxiliary injection had to be applied using existing access openings. There may be more optimal locations and this needs to be investigated however, the tests show that significant reductions in NOx can be obtained without modification to an operating furnace.

CROSS-FIRED INDUSTRIAL FURNACE TESTING

The furnace chosen for these tests is 600 t/d Float glass furnace using dual impulse side of port neck burners. The tests on the cross-fired furnace are expected to be carried out in November 2014. The basic testing procedure and measurements will be the same as for the End port furnace except that the injection must be applied to each port. A single port trial will be carried out before converting the remaining ports.

Although testing had already been carried out on the GDF SUEZ Cross fired2MW pilot furnace that simulates a single port, it was important to understand the effect of the interaction between ports of a multiport furnace. To this end an Acid/Alkali model of the cross-fired furnace to be tested was built. This revealed the criticality of location and operating parameters of the Auxiliary Injectors on this type of furnace. It also suggested the changes that may be required to the operation of the side of port neck burners, which may be required when Auxiliary injection is applied.

FURTHER DEVELOPMENT

Although the process will have been tested on two specific designs of furnace the range of furnace designs and sizes used by the glass industry is large therefore it would be helpful to have a method of determining the optimum Auxiliary injector location and performance of this process on a range of furnaces sizes and designs.

As has already been stated, full CFD modelling is possible but very time consuming, especially when many geometries and conditions must be tested. It is therefore proposed to develop a method of predicting the performance of furnaces using this firing process using zone modelling [3][3]. This technique allows a large number of options to be evaluated rapidly especially when there is good similarity between furnaces. Using this method a number of different furnaces will be simulated to allow rapid prediction of the Auxiliary injector requirements and performance for the most common furnaces used by the industry. No doubt there will be cases of furnace designs or operating conditions, which will require more detailed study by the methods referred to in this paper.

CONCLUSIONS

The work carried out so far has indicated that the Auxiliary combustion firing technique combined with conventional firing should be capable of reducing NOx on most conventional end-fired container furnaces to 500 mg/m^3(n) (@8%O$_2$) without loss of production, glass quality, efficiency or damage to the furnace. Although no Industrial testing has been carried out on cross-fired furnaces, pilot testing suggests similar results will be obtained on cross-fired furnaces. The location and operating conditions of the auxiliary injector is critical and must be optimised with the conventional firing conditions. Modelling in some form is required to determine these factors for any particular furnace.

The process can be applied to existing furnaces without modification to the furnace or if so only minor changes. However further studies may reveal that there are more optimal configurations that would require furnace changes that could only be applied at a repair or to a new furnace.

The equipment required is simple and existing controls can be modified to accommodate the Auxiliary gas so capital costs are low. Indications are that there will be no energy penalty and possible an improvement in efficiency. It may be that the process can be optimised for energy saving and NOx reduction. The method also eliminates the additional energy and CO$_2$ emissions associated with almost all post-furnace clean up techniques. Glass quality and pull rates do not appear to be affected by the process.

REFERENCES
1. Giesse, A. and Fleischmann, B. "EnergieeinsparungdurchVerbesserung des direktenWarme-eintrages an regenerativbefeuertenGlassschmelzwannen", Gaswarme International (58) No.5/2009.
2. Rhine, J.M. and Tucker, R.J. Modelling of Gas-fired Furnaces and Boilers, McGraw-Hill Books, 1991.
3. Tucker, R.J. et al, "Developments in the Application of Zone Modelling for Furnace Efficiency Improvements", British-French Flame Days, Lille, March 1990.

ENVIRONMENT AND ENERGY: FLUE GAS TREATMENT AND PRODUCTION OF ELECTRICAL POWER IN THE GLASS INDUSTRY

Alessandro Monteforte and Francesco Zatti
Area Impianti SpA
Albignasego, Padova ITALY

ABSTRACT

Flue gas treatment in glass industry is becoming more involved and elaborate to meet evolving environmental emission requirements. Traditionally, three compulsory systems are available to abate pollutants and acids (dusts, NOx, SOx): Electrostatic Precipitators, Bag Filters, and Ceramic Candle Filters eventually connected to an SCR reactor. Additionally, flue gas treatment specialists are trying to develop new technology solutions to optimize and increase operating and environmental performance, even using the so called "Lost Third" of Energy, exiting the melters with glass furnace combustion gases. Two typologies of heat recovery, thermal and electric, are generally possible. Cogeneration is also a third, mixed opportunity. The Rankine cycle is normally used to produce power by means of either water steam generation (direct exchange with flue gases) or organic fluid (indirect exchange, using thermal oil heated by flue gases). Where reliability and low maintenance are basic decision elements, Organic Rankine Cycles (ORC) are the most common choice and probably the best available technology. New generation organic fluid turbines are becoming more efficient and same-case comparisons show that yearly power production with ORC is at least as high as with traditional steam cycle, with the advantage of lower investment in manpower and time spent on this "appendix" of the glass furnace. The subject of this paper is the evaluation of the most reliable solution in terms of heat recovery linked with flue gas treatment (FGT), related to a specific plant situation, and a deep analysis of a typical heat path, considering technical and economical aspects.

INTRODUCTION

The use of the so called "Lost Third" of energy coming out from a glass furnace with the flue gases has been discussed for many years, without univocal answers. Moreover, the goal of achieving maximum energy efficiency must co-exist today with the need for a compulsory flue gas cleaning process that has its own requirements and limits. The goal of this paper is to describe the possibilities for heat recovery and energy generation, taking into account the different flue gas cleaning functions, and providing an overview of historical as well as most recent technical solutions for de-dust, de-acidification and de-NOx operations.

THE HEAT RECOVERY NEEDS

Two typologies of heat recovery are generally possible: thermal and electric; cogeneration is also a third, mixed opportunity. Since thermal heat recovery strictly depends on the use of heat, the thermo-dynamic related to it must be adapted and a general discussion cannot be easily applied. Some parameters discussed here can also be useful for a discussion on thermal heat recovery. The increasing cost of electricity as well as often available government incentives drives increased focus on heat recovery with the aim of power generation.

The Rankine cycle is used to produce power, normally realized by means of either water steam generation through direct exchange with flue gases or organic fluid through indirect exchange, using thermal oil heated by the flue gases. In western countries, where reliability and low maintenance are basic decision elements, Organic Rankine cycles (ORC) is the most common choice and probably the best available technology. New generation organic fluid turbine are more and more efficient and comparisons on the same case show that yearly power production with ORC is at least as high as with traditional steam cycle, with the advantage of lower investment in manpower and time spent on this "appendix" of the glass furnace.

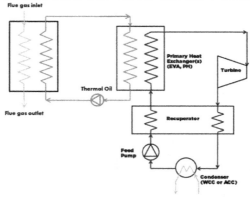

Figure 1. Typical organic Rankine cycle scheme

One of main weak points of the steam cycle is the strong dependence of its efficiency on steam temperature. A small reduction of flue gas temperature (always possible in the life of a glass furnace) translates to big losses of net produced power. This is not directly effective for the ORC system but nevertheless a steady thermal oil high temperature is also a good way to reach a high efficiency in the ORC. Temperature is a crucial point in designing efficient power produced by heat recovery, but these temperatures are strongly influenced by the flue gas treatment units.

THE FLUE GAS TREATMENT TEMPERATURE NEEDS
The main unit operations of a modern flue gas cleaning system are de-dusting, de-acidification, and de-NOx. De-dusting units are present in probably 90% of the flue gas furnaces in Europe, with a large presence of classical technology, including use of both electrostatic precipitators (ESP) and bag filters (BF). Any implementation of a heat recovery will need to cope with an existing or projected filtration system, and the two types have very different temperature ranges of application:
- ESP can be conveniently applied between 480 to 750°F (250 to 400°C)
- Bag filter can be applied at temperatures below 430°F (220°C), even if modern bags can bear peaks up to 500°F (260°C)

A third, partially proven technology can also be applied: the ceramic candles. Requirements of this particular technology will not be discussed in the present article.

Another key element is de-acidification, generally obtained with dry injection of a solid reagent: calcium or sodium based.
- Standard hydrated lime injection reaches a good efficiency at T>700°F (375°C)
- Sodium bicarbonate and sodium carbonate reach (at different levels) a good efficiency at T>355°F (180°C). They can be used in a wide range of temperatures but they involve some risk of sticky salt formation around 570-660°F (300-350°C)
- High specific hydrated lime can also be used, with good efficiency even in the typical sodium bicarbonate window

DeNOx SCR (Selective Catalytic Reduction) technology, even if proven and now well spread in the glass world, is not as well known as the other technologies. Particularly, almost 100% of the European container glass furnaces have no NOx treatment units. The main temperature problem of the SCR technology is poisoning of the catalyst linked to condensation of ammonium sulfate (called AS or ABS) salts that takes place in the micro-pores of the catalyst. This condensation clogs the micro-pores and therefore reducing its activity faster than normally expected. An SCR operating below the dew point of ammonium sulfate is called "low T SCR", and a unit operating at T higher than the dew point is called "high T SCR". Operation at "low T" is possible but requires some specific precautions. Empoisoning by ammonium sulfate is a reversible process because, with a temporary increase of T, salts can be evaporated and the catalyst completely regenerated.

A generic assumption is that the minimum operating temperature for an SCR is 570-590°F (300-310°C). Temperatures lower than 300°C can, however, still be called "high" and operations at low T are also possible. The main requirement is to clearly understand the sulfate dew point. The latter depends on the ammonia and SO_3 concentration in the flue gas and so strictly on the de-acidification process done upstream. A good de-acidification process can not only reduce the SOx, but also reduces the ratio between SO_3 and SO_2 from about 10% down to levels around 0.25% to 0.5%.

Part of the SO_2 is also converted to SO_3 by the catalyst itself; this quantity depends on the type of catalyst used and from the operating temperature. Generally speaking, in a low-T SCR oxidation of SO_2 is almost negligible, when above 300°C the conversion can reach values of 0.5%.

Another important element in the SO_3 conversion is fuel: oil fired furnaces will have not only higher SOx levels, but also higher SO_3/SO_2 ratios because of the heavy metals present in the fuel oil, catalyzing the oxidation of SO_2 into SO_3.

Also, the sulfate dew point is not reached not at a single temperature but within a range of temperatures. The condensation of these salts occurs at higher temperatures in the catalyst micro-pores and lower temperatures in bulk.

Once the dew point range is defined, we can decide if we want to operate at high temperature or at low temperature. In this second case, it is necessary to estimate the time of deactivation of the catalyst. This will depend on the actual concentration of SO_3, as previously mentioned, and the difference from the dew point and the operating temperature. The higher the difference between dew point and operating temperature, the higher will be the frequency of the thermal regenerations required to remove the ammonium sulfates. It's clear that if this frequency is too high, it would be wiser to decide for a high temperature application.

Thanks to a system patented by Area Impianti and used for 10 years now, regeneration can be done on-line, automatically and with a negligible use of energy. In fact this type of SCR is conceived with metal casing divided into compartments, which can be sectioned by means of sealing valves, allowing module by-module regeneration. Benefits of this system are: gas consumption reduction for regeneration, emissions below limits also during regenerations, on-line maintenance is possible. Though it's clear that an SCR conceived for low T application is more complex and therefore slightly more expensive than a high T application, it can solve big problems on existing plants, bringing an overall project cost saving.

As an example, we can say that an SCR applied to a container glass natural gas fired furnace with SOx reduction from 800 to 300 mg/Nm^3 through sodium bicarbonate and content of 1200 mg/Nm^3 of NOx to be reduced to 500 mg/Nm^3 can operate at 430°F (220°C) with approximately one regeneration every 40 days.

Furthermore, an SCR applied to a float glass natural gas fired furnace with SOx reduction from 1000 to 300 mg/Nm³ and content of 2000 mg/Nm³ of NOx to be reduced to 500 mg/Nm³ can operate at 520°F (270°C) without any need for regeneration.

CORROSION AND STACK PROTECTION

Calculating SOx levels will help to define the minimum temperature at the end of the treatment/heat recovery line. Corrosion problems in the heat recovery units, and also at the stack, must be clearly taken into account.

For a standard natural gas fired furnace, we can assume that minimum T of the metallic surfaces should be 355°F (180°C) before de-acidification and 300°F (150°C) after treatment. Staying on the safe side, the flue gas temperatures at ID fan outlet are generally kept above 355°F (180°C), also to avoid problems of re-condensation of ammonium sulfates after the SCR.

TYPICAL HR+FGT LINES

Following the above mentioned indications, two example lines with integrated heat recovery and complete flue gas cleaning systems are presented.

a) One float natural gas fired furnace, 830 US ton/d (750 t/d), 3,000,000 SCFH (85.000 Nm³/h) @ 840°F (450°C).

Line is composed by:
- 1st step heat recovery down to 620°F (325°C)
- Injection of hydrated lime for de-acidification
- ESP for de-dusting
- Hi-T SCR at 590°F (310°C)
- 2nd step heat recovery down to 355°F (180°C)
- ID fan and stack

Total heat recovered: 9.1 MWth
Gross power production: 2.2 MWel
Typical pay-back (Italy): 2-2.5 years

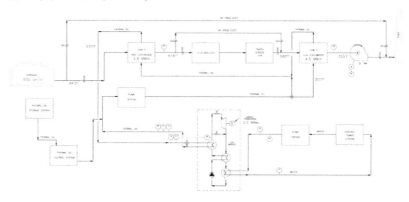

Figure 2. FGT system and ORC applied to single float furnace with 830 ton/d

b) Two container natural gas fired furnaces, 275+275 US ton/d (250+250 t/d), 1,600,000 SCFH (45.000 Nm³/h) @ 930°F (500°C).

Line is composed by:

- heat recovery down to 430°F (220°C)
- Injection of sodium bicarbonate for de-acidification
- Bag filter for de-dusting
- Low-T SCR, (with automatic regeneration system) at 410°F (210°C)
- ID fan and stack

Total heat recovered: 5.1 MWth
Gross power production: 1.2 MWel
Typical pay-back (Italy): 3-3.5 years

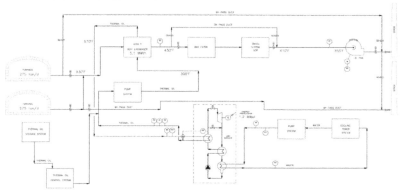

Figure 3. FGT system and ORC applied to 2 container furnaces with 275 ton/d

CONCLUSIONS
1. Heat recovery and flue gas treatment can coexist profitably and more specifically:
   - SCR and heat recovery are compatible
   - Bag filters and heat recovery are compatible, now and in the future when SCR will be applied.
2. Treatment of flue gases and heat recovery must be designed by experts in both fields because the interactions are crucial for the optimization of both heat recovery systems and good and reliable operation of flue gas cleaning system.

A look to the future is necessary today when designing the heat recovery in case an SCR may be needed in the future or lower pollutant limits will be required or a flue gas cleaning system in case future heat recovery will be implemented. That's probably the real meaning of "clean energy".

OPTIMELT™ REGENERATIVE THERMO-CHEMICAL HEAT RECOVERY FOR OXY-FUEL GLASS FURNACES

A. Gonzalez and E. Solorzano
Grupo Pavisa, S.A. de C.V., Naucalpan, Edo. México

C. Lagos and G. Lugo
Praxair México S de RL de CV, Col. San Salvador Xochimanca, México D.F.

S. Laux, K.T. Wu, R.L. Bell, A. Francis, and H. Kobayashi
Praxair, Inc., Danbury, CT, USA

ABSTRACT

The operation of glass furnaces with oxy-fuel combustion in combination with advanced heat recovery is a compelling low cost solution. Praxair has developed a new heat recovery technology for oxy-fuel fired furnaces in which regenerators are used in a similar way as conventional regenerators, which recover waste heat from the furnace flue gas. The OPTIMELT™ Thermo-Chemical Regenerator (TCR) technology uses the stored waste heat to reform a mixture of natural gas and recirculated flue gas to hot syngas, resulting in efficient heat recovery. The natural gas reacts endothermically in the hot checker pack with the water vapor and CO_2 in the recycled flue gas, forming H_2 and CO as a hot syngas fuel. The OPTIMELT™ TCR system is simple and operated at atmospheric pressure without the need of catalysts or separate steam generation. It reduces fuel consumption of an oxy-fuel fired furnace by about 20% and offers an attractive conversion option for existing air-regenerator furnaces, with more than 30% fuel reduction compared to the air-fuel base case. This technology is currently being demonstrated on a 50 t/d commercial glass furnace at Pavisa. This paper introduces the operating principles of the TCR and highlights some milestones recently achieved at the demonstration site. Initial data from the TCR operation at the host furnace are presented and discussed.

INTRODUCTION

Oxy-fuel glass melting was first developed for large glass melting furnaces in early 1990s in the US. At that time reduction of NOx emission was a primary driver and 80 to 90% reduction in NOx emissions was demonstrated. The fuel consumption by oxy-fuel glass melting at the glass plant level is generally reduced by 10-20% compared to those from regenerative furnaces. However, these fuel savings typically do not outweigh the cost of oxygen. As a result, without a strong NOx reduction driver, state of the art regenerative air-fuel furnaces remain the choice of the industry.

Reducing energy consumption from glass melting operation has been a continuous goal for the glass industry, especially for glass plants located at high priced natural gas geographic regions. In recent years many improvements and trials regarding flue gas heat recovery have been reported [1, 2]. In oxy-fuel fired glass furnaces about 1/4 of the fuel energy input is lost as sensible heat of flue gas. The operation of glass furnaces with oxy-fuel combustion in combination with heat recovery is a compelling approach to significantly reduce the energy consumption of oxy-fuel glass furnaces, while still taking advantage of the benefits of oxy-fuel glass melting including the substantial reduction in the flue gas volume and reduced emissions. In the state of the art air heating regenerator, about 65% of the waste heat in the flue gas is recovered by preheating combustion air. If the same technology is applied to recover waste heat from an oxy-fuel fired glass furnace by preheating the combustion oxygen, then only about 33%

of the waste heat is recoverable due to large differences in heat capacity rates between the flue gas and the oxygen streams.

This paper presents information on the newly developed OPTIMELT™ heat recovery system for oxy-fuel furnaces which integrates checker-packed regenerators and flue gas recirculation, with an oxyfuel furnace. This technology is presently being field demonstrated on a 50 t/d commercial glass furnace. The concept and advancement of the OPTIMELT™ technology has been disclosed [3] and described [4] previously. The OPTIMELT™ heat recovery technology has potentials for fuel savings of about 20% compared to oxy fuel and 30% compared to air-regenerator furnaces.

OPTIMELT™ HEAT RECOVERY PROCESS

The heart of the technology is based on a unique waste heat recovery concept called "Thermo-Chemical Regenerator" (TCR) in which heat stored in a regenerator checker during the exhausting/heating cycle is recovered during the preheating/reforming cycle by preheating and reforming a mixture of natural gas and recycled flue gas. Figure 1 shows this cyclic heat recovery process.

The heating cycle is similar to the conventional regenerator heating cycle in which flue gas waste heat is transferred to and stored in the checker. The unique feature of the TCR process occurs during the reforming cycle where a portion of the cooled flue gas is recycled back (Recycled Flue Gas, RFG) to the bottom of an already preheated regenerator and mixed with a reforming fuel (e.g., CH_4). The RFG/CH_4 mixture is heated by the hot checker. When the gas mixture is heated above a certain temperature various endothermic chemical reactions occur. For example, CH_4 is reformed by CO_2 and H_2O in the RFG to form CO and H_2. The reformed gas or "syngas" is combusted with oxygen in the glass furnace, thus providing thermal energy for glass melting. The problem of using heat recovery for oxy-fuel combustion flue gas is solved by using endothermic chemical reactions to eliminate the imbalance of heat capacity ratios.

The OPTIMELT™ TCR process takes advantage of the high H_2O and CO_2 concentrations (~85% total) in the flue gas of oxy-fuel combustion and synergistically utilizes them as reactants for the endothermic reforming reactions. By comparison, the combined H_2O and CO_2 concentration of the air combustion flue gas is only about 30% of the flue gas and the remaining 70% N_2 cannot participate in the endothermic reactions.

The operation of TCR is very similar to the conventional regenerators and the volume of the checkers is only about 1/3 of that of the conventional air regenerators used in the same size furnace. The reforming reactions proceed at atmospheric pressure without the need of catalysts or separate steam generation.

Fuel savings by the OPTIMELT™ system vary depending on furnace and glass types. As an example, Figure 2 shows a calculation of the fuel savings for a 300 t/d container furnace with 500 kW boost at 50% cullet over a furnace life of 12 years. The furnace ageing factors (increase in fuel consumption over time) were calculated to be 1.35, 0.54 and 0.71% per year for the air fired furnace, the oxy-fuel furnace and the OPTIMELT furnace, respectively, based on a set of consistent assumptions on increases in wall heat losses, air infiltration and deterioration in regenerator performance. The fuel savings at mid-campaign by the OPTIMELT™ system is about 28% relative to the state of the art air regenerative furnace in this container furnace example. For larger units such as flat glass furnaces, fuel savings by the OPTIMELT™ technology are expected to be higher because the total wall losses are lower per unit glass pulled thus more flue gas heat can be recovered.

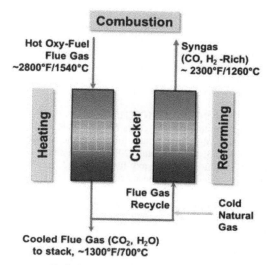

Endothermic Reforming Reactions:

$$CH_4 + H_2O \rightarrow CO + 3H_2 \quad 9.1 \text{ MJ/Nm}^3 \text{ CH}_4 \text{ (229 Btu/scf-CH}_4\text{)}$$
$$CH_4 + CO_2 \rightarrow 2CO + 2H_2 \quad 11 \text{ MJ/Nm}^3 \text{ CH}_4 \text{ (275 Btu/scf-CH}_4\text{)}$$

Figure 1: OPTIMELT™ thermo-chemical heat recovery with regenerators

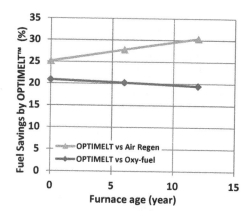

Figure 2: Example for fuel savings from OPITMELT™ technology vs. furnace age

SUMMARY OF PREVIOUS DEVELOPMENT EFFORTS

Praxair's OPTIMELT™ TCR heat recovery technology has been under development since 2010. Development stages for the technology included bench and pilot scale TCR tests, syngas burner development, process safety review, and extensive computer modeling related to reforming chemistry, glass furnace flame coverage and performance, and flow and heat transfer in TCR regenerators. These technology development milestones that led to the current field demonstration have been described previously [5].

DESCRIPTION OF THE DEMONSTRATION SITE

The demonstration site is located at Pavisa's Furnace 13 in Mexico. Pavisa manufactures glass and crystal products for global wine, liquor, food, perfume, and pharmaceutical industries. Furnace 13 has a melting area of approximately 29 m^2 with a single charger on the left and a nominal pull rate of 50 t/d. The furnace is oxy-fuel fired and has three Praxair Wide Flame Burners on each breast wall placed in an opposed configuration. The furnace produces various flint and color glasses with frequent color changes.

The addition of the TCR heat recovery system to the furnace changed the combustion from the side-port firing of the oxy-fuel burners to end-port firing of the syngas burners. In anticipation of the need for optimizing the syngas flame's characteristics and coverage during the process startup phase, four thermocouples (TC) were installed on the furnace's right breast wall and one on the left. These thermocouples are spaced along the length of the walls at the same height from the glass surface. In addition, a video camera was mounted to the furnace's front wall for recording in-furnace batch and flame movements. We have found that the combination of these additional TCs and camera has provided valuable operational information, insight, and documentation for the first startup of the OPTIMELT™ heat recovery process.

Figure 3 shows a pictorial view of the TCR installation at the demonstration site. The two TCR regenerators had to be retrofitted into a narrow space between a building wall and the furnace rear wall. The modular regenerator steel shells are reinforced for seismic loading. They were brought individually to the location and lifted from the ground into steel columns at the corners as crane access was not possible at the site. Placed between the two regenerator structures is the flue gas recirculation skid. It was prefabricated in three modules and put into place after the left regenerator shell was completed. The port necks are downward sloped in order to gain space for additional checker layers for heat recovery.

The existing oxy fuel burners were retained and the furnace can return to regular oxy-fuel firing at any point of operation. The stack also remained in place and a gate damper was installed between the stack and the original furnace exhaust port to block the flow of flue gas into the stack for TCR operation. The TCR system exhausts the cooled flue gas directly into the bottom of the existing stack. It is worthwhile to note that the furnace can operate with a mixed firing mode, i.e. TCR syngas firing combined with partial oxy burner firing.

SITE CONSTRUCTION AND PREPARATION

Integration of the two TCR regenerators, port necks, flow control skids, and connecting pipes at the Pavisa host site started in May 2014 during a 17-day shutdown originally scheduled for furnace cold repairs. Before the furnace shut down, many TCR components were prefabricated in modular sections and installed. The confined installation location required careful staging of the work and detailed coordination of the multiple contractors. The furnace was back to oxy fuel firing for glass production in early June. The installation of all OPTIMELT™ system components including controls and instruments were completed at the end of June. Extensive pre-startup safety reviews and training classes were conducted jointly with the host site in the first half of July.

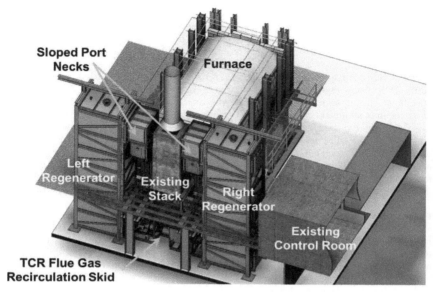

Figure 3: Overview of the TCR installation at Pavisa

INSTALLATION OF REFRACTORY CORROSION SAMPLES

Glass furnaces with producer gas (PG) fired regenerators have been operated successfully for many years in the last century by the US glass industry, and are still popular in certain geographic regions. As part of the development work the smooth operation of a 750 metric t/d float furnace using PG fired regenerators, and several other smaller PG-fired furnaces of less than 100t/d were observed in China in 2012. Although producer gas is heated up to 1200-1300°C and contains similar components such as high concentrations of CO and H_2 as reformed TCR syngas, and is commercially produced using conventional checkers, concerns about checker and refractory life exist. The authors are not aware that any long-term test data in the public domain exists which can be used to predict the longevity of checker or refractory life under relevant reforming conditions.

Therefore, the two TCR regenerators were designed to be able to house up to 130 refractory and checker samples at different regenerator elevations so that these samples can be exposed to a wide range of temperature and gaseous conditions. Before the regenerator heat up, about 100 samples were placed into the designed sample holders at locations which were still accessible after the construction. The plan is to let the samples stay at their respective locations for at least six months until they are retrieved for analysis. It is expected that very useful and practical refractory and checker performance data will be generated for commercial OPTIMELT™ implementation.

TCR REGENERATOR HEAT-UP AND PROCESS STARTUP

Heating up the TCR regenerators began in the second half of July. The regenerators were firstly being heated up simultaneously by hot flue gases from the furnace in oxy-burner firing, at controlled refractory heat-up rates according to refractory specifications used in the regenerators.

Once the regenerator temperatures were above a minimum, heat up was switched to cyclic mode where the regenerators were brought up to operating temperatures by hot flue gases directed to each regenerator alternatively. During the heat up process, the furnace by-pass damper was controlled automatically to maintain the furnace pressure and the amount of hot flue gas going into the regenerators.

In early August, several short term reforming tests were concluded at 50%, 66%, and 100% TCR firing rate. Syngas burner flames were observed and adjusted to have good operational characteristics. When the TCR was partial firing for example at 50%, the remaining of the required total fuel for glass melting was directed to active oxy fuel burners.

The startup trials also demonstrated that furnace pressure can be controlled to a very precise level under all three firing conditions, namely, oxy fuel burner only, TCR syngas burner only, and TCR-oxy fuel burner mixed firing. First temperature and flow data were collected and analyzed. In addition, detailed mapping of regenerator and port neck surface and ambient temperatures were concluded. These temperature data will be used to calculate convective wall heat losses appropriately for total energy balances later.

PROCESS OPTIMIZATION

The process startup phase was also used to improve on several hardware and software items, which is typical for such a complex technology implementation. However, at no time was there any interruption of high quality glass production. Although the nature of these items was non-critical, nevertheless they required corrective action before longer term trials could be commenced. The timing of the actuators which control the injection of various gases was also adjusted, based on data recorded and the vast amount of previous experience gained from pilot scale TCR tests.

As an example of the OPTIMELT™ TCR operation, Figure 4 shows the cyclic changes of the temperatures at the top and bottom of the regenerators for about 4 cycles. After the start of the reforming on the left side the temperatures in this regenerator drop due to the endothermic reactions of the natural gas with the recirculated flue gas. At the same time the temperatures on the right side increase as hot flue gas heats the checkers. The cycle repeats every 20 min. Optimized cycle times depend on thermally balancing both side regenerators over a longer time.

It is expected that the furnace will be operating under 100% TCR syngas firing for long-term thermal performance evaluation starting in the fourth quarter of 2014. The first application of the technology in a commercial environment so far has shown that the response of the temperatures in the furnace are as expected and that the flames can be properly shaped to meet the requirements of the furnace. The furnace pressure can be easily controlled under syngas or syngas-oxy fuel mixed firing configurations.

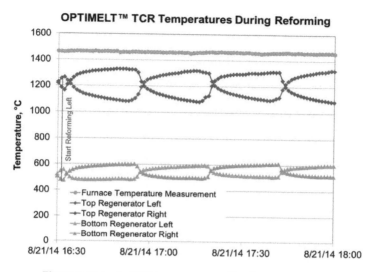

Figure 4: Example TCR temperature profiles during initial trial

SUMMARY AND FURTHER WORK

Praxair's OPTIMELT™ Thermochemical Regenerator (TCR) technology is an important step towards higher energy efficiency of oxy-fuel fired glass furnaces. The estimated energy consumption is reduced by approximately 20% versus oxy-fuel combustion and 30% versus state of the art regenerative air furnaces making OPTIMELT™ TCR a compelling choice for future conversions of air furnaces. Naturally, associated with the reduction in fuel consumption is a corresponding reduction of combustion CO_2 emissions.

In principle the TCR technology is based on oxy-fuel combustion with similar expected low NOx, CO, SO_2 and particulate emission levels than oxy-fuel furnaces that produce similar glass. Another inherent oxy-fuel advantage is the smaller cooled flue gas volume and associated handling equipment.

The OPTIMELT™ Thermochemical Regenerator (TCR) technology has been successfully started up at commercial scale at Pavisa in Mexico. Pavisa and Praxair have very closely collaborated to integrate the technology into the existing 50 t/d furnace under challenging circumstances. The optimization of the OPTIMELT™ heat recovery technology at the demonstration site is continuing. The initial operational data and fuel savings are very encouraging and a detailed quantification is in progress as of the time of this writing. Praxair expects that a 300 tpd size OPTIMELT™ TCR system will be ready for commercial application in 2015.

Near and longer term goals for this new technology demonstration include the following:
• Optimize syngas burner flames for different types of glass productions
• Investigate the effect of process parameters on heat recovery efficiency
• Achieve long-term operational record
• Analyze long term performance data to determine energy savings of the host furnace

- Develop and finalize commercial TCR system maintenance schedules
- Retrieve and analyze refractory and checker corrosion samples from the TCR regenerators

REFERENCES

1. Zippe, P., "Recent Development of Batch and Cullet Preheating in Europe – Practical Experience and Implications", pp. 3-18, 71st Glass Problems Conference, Oct., 2011.
2. Limpt, H., and Beerkens, R., "Energy Recovery from Waste Heat in the Glass Industry and Thermochemical Recuperator", pp. 3-15, 73rd Glass Problems Conf., Oct., 2013.
3. Kobayashi, H., US Pat. 6,113,874, Sep. 5, 2000.
4. Kobayashi, H., Wu, KT, Bell, R.L., "Thermochemical Regenerator: A High Efficiency Heat Recovery System for Oxy-Fired Glass Furnaces", DGG/AcerS Conference, Aachen, May 28, 2014.
5. Gonzalez A., and Solorzano, E., et. al., "Commercial Demonstration of OPTIMELT™ Thermo-chemical Heat Recovery for Oxy-Fuel Glass Furnaces", 29th A.T.I.V. Conference and 12th European Society of Glass Conference, September 21-24, 2014, Parma, Italy.

Refractories

WEAR OF BASIC REFRACTORIES IN GLASS TANK REGENERATORS

David J. Michael, H. Edward Wolfe, and Laura A. Lowe
North American Refractories Company
Pittsburgh, PA 15108, USA

ABSTRACT

This paper reviews the chemical reactivity of basic refractories used in glass tank regenerators with V_2O_5 in oil-fired furnaces, with sodium sulfate and with alkali. The rationale for using magnesia brick with high levels of forsterite in situations where V_2O_5 attack is anticipated is given. A post-mortem examination showing depletion of CaO in basic brick exposed to sodium sulfate is shown. Laboratory work on reactivity of basic brick with sodium sulfate is also presented to show the effect of CaO depletion on strength of basic refractories. Alkali attack on forsterite-bonded brick is discussed. A new type of forsterite-bonded brick containing no ZrO_2 with properties similar to those of magnesia-zircon brick but with substantially better alkali resistance is introduced.

INTRODUCTION

Refractory suppliers and users make judgments about what refractories to use in a given application through a combination of studying the environment of the application, observing the performance of refractory products, conducting simulative testing, and doing post-mortem examinations of refractories. Determining the appropriate products to use in glass tank regenerators is one of the more challenging refractory issues due to the long-term, cyclic nature of the application, where refractories are required to last 10 years or longer. This paper reviews some of the principal wear mechanisms of basic brick used in glass tank regenerators to aid in the selection of appropriate refractories.

LITERATURE REVIEW OF V_2O_5 ATTACK ON BASIC REFRACTORIES IN GLASS TANK REGENERATORS

Considerable academic work has been done on attack of basic refractories by V_2O_5. Eke and Brett [1] made the following statement: "The principal component of heavy fuel oil ash is vanadium pentoxide, V_2O_5, amounting to ~0.04%, which reacts with basic refractories by attacking the silicate bond and consequently reducing the high temperature mechanical properties of the brick. Examination of magnesite bricks from the top course of a regenerator showed an accumulation of V_2O_5 with contents ranging from 1-2% and evidence of considerable migration of lime within the bricks during service."

In understanding how V_2O_5 reacts with basic brick, one focuses on the following factors:

(1) The temperatures at which the silicate minerals in burned magnesia brick begin to melt
(2) The sequence of reactions that occur due to V_2O_5 deposition in a brick containing free magnesia
(3) The amount of bond phase in the brick

All burned magnesia brick contain CaO and SiO_2 as impurities. Table 1 illustrates the mineral phases that exist at equilibrium in a burned magnesia brick at various lime-to-silica ratios; the temperatures of initial liquid formation of those minerals are included in the table.

123

The temperature of initial liquid formation in basic refractories is important because the mechanical properties of the brick deteriorate once liquid begins to form.

Table 1. Mineralogical composition of silicate bond phases in burned magnesia regenerator brick			
Mineral	Composition	Lime-to-Silica Ratio (weight basis)	Temperature of Initial Liquid Formation, °C
Tricalcium Silicate	$3CaO \cdot SiO_2$	2.80	1925
Dicalcium Silicate	$2CaO \cdot SiO_2$	1.87	1925
Merwinite	$3CaO \cdot MgO \cdot 2SiO_2$	1.4	1575
Monticellite	$CaO \cdot MgO \cdot SiO_2$	0.93	1500
Forsterite	$2MgO \cdot SiO_2$	0	1850

With the higher lime-to-silica ratios above 1.87 liquid does not form in a burned magnesia brick until very high temperatures. At intermediate lime-to-silica ratios of 1.4 and 0.93 liquid forms at substantially lower temperatures. If the lime-to-silica ratio were zero, i.e. if there were no CaO in the composition, the silicate phase would be forsterite, and the temperature of initial liquid formation would again be quite high. Since all commercial magnesia contains some amount of CaO, having a lime-to-silica ratio of zero is not possible, and in brick with lime-to-silica ratios below 0.93 there will be a combination of monticellite and forsterite. In addition, all commercial magnesias contain small amounts of Al_2O_3, Fe_2O_3 and B_2O_3 that will lower the temperatures of initial liquid formation given in the table above. Also, in some cases having a combination of minerals will lower the temperature of initial liquid formation. For example, for lime-to-silica ratios between 1.87 and 1.4, the temperature of initial liquid formation is 1400°C, which is 175°C lower than the temperature of initial liquid formation of the lowest melting endpoint mineral, merwinite [2].

One issue with infiltration of V_2O_5 into a burned magnesia brick is that it will react with the lime-containing silicate minerals to form tricalcium vanadate, $3CaO \cdot V_2O_5$, which melts incongruently at the relatively low temperature of 1380°C. In a burned magnesia brick with a dicalcium silicate bond phase, the following series of reactions would occur with increasing amounts of V_2O_5[3]:

(1) The V_2O_5 would take CaO away from the dicalcium silicate to form tricalcium vanadate, $3CaO \cdot V_2O_5$. This would reduce the amount of CaO available to react with the SiO_2 in the brick, effectively lowering the lime-to-silica ratio of the silicate phase; some of the dicalcium silicate would be converted to merwinite. If enough V_2O_5 were introduced, all of the dicalcium silicate would be consumed, and the silicate would become the relatively low-melting merwinite phase.

(2) If V_2O_5 deposition continued, additional tricalcium vanadate would form, further lowering the lime-to-silica ratio of the silicate phase until all the merwinite was consumed and the only silicate phase was monticellite.

(3) With additional deposition of V_2O_5, the monticellite would begin to be consumed and eventually only tricalcium vanadate and forsterite would be present.

(4) Forsterite itself can be altered by V_2O_5 to form a garnet phase that exhibits liquid formation.

If a brick with a high-lime-to-silica ratio were placed into a glass tank checker in an oil-fired furnace where V_2O_5 deposition occurred, there would be a progression of reactions resulting in formation of low-melting phases that would contribute to deterioration of the strength and creep resistance of the brick. This would be influenced by the total amount of CaO and SiO_2 in the brick.

Current practice is to use forsterite-bonded magnesia-zircon brick in condensate zones of regenerators when V_2O_5 attack caused by oil firing may occur. The systems MgO-2MgO·SiO_2-V_2O_5 and MgO-2CaO·SiO_2-V_2O_5 have been studied in detail [4,5]. The researchers who studied those systems recommended a brick with the following chemistry for use in glass tank regenerators when V_2O_5 attack may occur:

MgO	90.0
CaO	<1.0
SiO_2	8.2
Fe_2O_3	<0.4
Al_2O_3	<0.4

In basic brick, the bond phases tend to lie at grain boundaries, and it has been shown that mechanical properties deteriorate quickly when liquid forms. The bond phases are crystalline rather than glassy when examined at room temperature. The absence of glassy phases in basic brick implies that the liquids that form at elevated temperatures are quite fluid, and the fluidity is likely the reason that mechanical properties deteriorate so readily when liquid is formed.

With high purity burned magnesia brick, for example those above 98% MgO, all of the bond phase can become liquid with even a small amount of V_2O_5 deposition, regardless of whether the bond is dicalcium silicate or forsterite. This is illustrated by the phase diagrams in Figures 1 and 2. Figure 1 is the 1450°C isothermal section for the MgO-dicalcium silicate [6] - V_2O_5 system, and is an equilateral triangle with the apices representing the compositions of the three components. The location on the MgO-Ca_2SiO_4 line near the MgO apex represents the composition of a high purity dicalcium silicate-bonded burned magnesia brick. It is easy to see that slight additions of V_2O_5 to a composition composed of a high amount of MgO with a small amount of dicalcium silicate would cause all of the dicalcium silicate to become liquid. The path to the liquid state would follow the sequence of reactions described earlier in this paper.

Figure 2 shows that a similar situation would exist with a high purity magnesia composition bonded by forsterite [7]; just a small amount of V_2O_5 would cause the forsterite bond to become completely liquid. However, as the compositions on the $MgO-Mg_2SiO_4$ line moves towards higher levels of forsterite, the forsterite does not become completely liquid when small amounts of V_2O_5 are added. A brick with the chemistry above was proposed because when V_2O_5 is introduced into such a composition, even though liquid can potentially form, so much forsterite bond phase is present that even with significant V_2O_5 in the brick, substantial amounts of forsterite persist. In other words, although liquid can form when V_2O_5 deposits in a brick with a high amount of forsterite, all of the bond phase does not transform into a liquid as can happen with higher purity brick with less bond phase.

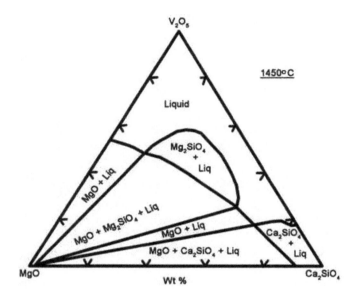

Figure 1. The System MgO-Dicalcium Silicate-V_2O_5 at 1450°C

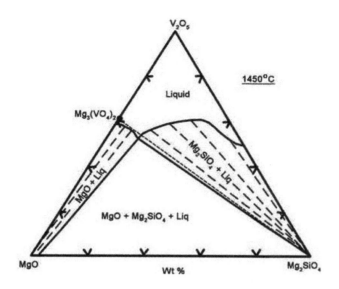

Figure 2. The System MgO-Forsterite-V_2O_5 at 1450°C

Current practice is to use magnesia-zircon brick that contain 14 to 19% forsterite in lower checkers when V_2O_5 attack is anticipated. The potential for liquid formation is dependent on many factors, including the concentration of V_2O_5 and the temperature at which the checker is operating. In addition, soda attack on forsterite can be a consideration under certain conditions, such as in reducing atmospheres.

REACTIVITY OF CAO IN THE BOND PHASE OF BASIC REGENERATOR REFRACTORIES IN THE PRESENCE OF SODIUM SULFATE

Much has been written about how V_2O_5 from fuel oil can react with CaO in the silicate minerals in basic brick to produce low melting phases that can increase the probability of checker failures by causing the hot strength and creep resistance of the brick to deteriorate. Less has been written about the reactivity of CaO in the bond phase of basic brick with sodium sulfate that is present in regenerators.

POST-MORTEM EXAMINATION AFTER USE IN GLASS TANK REGENERATORS

A major U.S. float glass manufacturer used a burned magnesia brick as part of the furnace regenerator package in a gas-fired furnace. After approximately 5 years' service, the checkers experienced a partial collapse. Several pieces of broken brick were pulled from the debris pile under the checkers for analysis. Table 2 contains chemical analysis data on a representative sample of the used brick, on the original unused brick, and of buildup taken from the surface of the brick. The unused brick had a CaO content of 2.05% while samples taken from the exterior and interior of the used brick had much lower CaO contents of 0.19 and 0.14%, indicating depletion of CaO during service. The buildup on the surface of the brick contained

2.59% CaO, suggesting that at least some of the CaO from the brick had migrated into the buildup.

The buildup contained mainly thenardite (Na_2SO_4), and anhydrite ($CaSO_4$). There were trace-to-minor amounts of periclase (MgO) and vanhoffite ($Na_6Mg[SO_4]_4$) in the buildup. While no magnesium sulfate was detected, the presence of vanhoffite suggested that some of the magnesia had entered the thenardite structure. It was suspected that depletion of CaO from the bond phase of the brick may have been a contributing factor in the partial checker collapse.

Table 2. Chemical analysis data on representative bricks				
	Buildup Taken from Exterior of a Used Regenerator Brick	Sample Cut from the Outer Surface of a Used Regenerator Brick	Sample Cut from the Interior of a Used Regenerator Brick	Chemistry of the Unused Regenerator Brick
	As-Received Basis	Calcined Basis	Calcined Basis	
Silica (SiO_2)	0.40%	0.64%	0.60%	0.76%
Alumina (Al_2O_3)	0.32	0.16	0.13	0.12
Titania (TiO_2)	<0.01	0.00	0.00	0.00
Iron Oxide (Fe_2O_3)	0.53	0.53	0.52	0.48
Zirconia (ZrO_2)	0.40	0.00	0.00	--
Lime (CaO)	2.59	0.19	0.14	2.05
Magnesia (MgO)	13.5	95.1	95.4	96.49
Manganese Oxide (MnO)	0.04	0.00	0.00	--
Phosphorous Pentoxide (P_2O_5)	0.06	0.01	0.01	--
Sulfur Trioxide (SO_3)	48.4	--	--	--
Soda (Na_2O)	33.1	3.27	3.09	--
Potash (K_2O)	0.08	0.00	0.00	--
Total:	99.42%	99.90%	99.89%	99.90%
		As-Received Basis	As-Received Basis	
Soda (Na_2O)	--	10.95	9.1	--
Sulfur Trioxide (SO_3)	--	15.4	13.2	--
X-Ray Diffraction				
Thenardite (Na2SO4)	Major	--	--	--
Anhydrite (CaSO4)	Minor	--	--	--
Vanhoffite (Na6Mg[SO4]4)	Trace-minor	--	--	--
Periclase (MgO)	Trace-minor	--	--	--

SIMULATIVE REGENERATOR TESTING

To better understand the wear mechanisms of basic regenerator brick in contact with sodium sulfate, testing was done to simulate the conditions in the condensate zone of a glass tank regenerator. This testing was based on a method first described by John LeBlanc in the Journal of the Canadian Ceramic Society in 1982 [8]. In the test, refractory brick samples measuring 19x19x45 mm were cycled between 732°C, where sodium sulfate is solid, and 1010°C, where sodium sulfate is liquid. (The melting point of Na_2SO_4 is 884°C.)

Eight sample bars were cut from the interior of each refractory and 4 each were placed inside two 94% Al_2O_3 containers measuring 10x6x3 inches. The container interiors were painted with a mixture of Cr_2O_3 and phosphoric acid to inhibit reaction between the samples and the containers. The sample bars were placed on end within the containers (19 x 45-mm faces parallel with the vertical).

One of the containers was designated "open" and the other was designated "closed." In the "open" container, enough Na_2SO_4 was added to cover the samples completely. The "closed" container did not contain any Na_2SO_4, only the samples. A lid was placed on top of the "closed" container in an attempt to isolate the samples from any Na_2SO_4 that might be present in the furnace atmosphere as a result of volatilization from the "open" container. The "closed" container was included so that there would be reference data for cycled samples that were not subject to chemical attack by Na_2SO_4. It was necessary to periodically replenish the Na_2SO_4 in the open container during the test period.

The samples were cycled between 732°C and 1010°C for two weeks, for a total of 125 cycles. After cycling, the furnace temperature was raised to 1177°C to simulate volatilization of Na_2SO_4 during a checker burnout. The samples were tested for % linear change, % weight change, and modulus of rupture at 1038°C.

The MOR was measured in 3-point bending using an Instron mechanical tester equipped with a small electric furnace. The cross-head speed was 0.25 mm/minute and the span was 32 mm.

RESULTS OF SIMULATIVE CHECKER TESTING

Table 3 contains the chemical analyses of the brick samples tested in the simulative checker test. A high-purity magnesia brick, two intermediate-purity magnesia brick, and a magnesia-zircon brick were tested.

Table 4 contains the % linear change data on the brick after testing. Linear changes were from slight expansion to slight shrinkage. The presence of sodium sulfate did not have a uniform effect on dimensional change during the test.

As illustrated in Table 5, all samples exhibited weight pickup during the test. The highest weight gains in all cases but one were for samples cycled in contact with sodium sulfate. Even the samples not in contact with sodium sulfate exhibited weight gains.

Table 3. Chemical analyses of brick tested in simulative checker testing				
	High Purity Burned Magnesia Brick	Intermediate-Purity Burned Magnesia Brick #1	Intermediate-Purity Burned Magnesia Brick #2	Magnesia-Zircon Brick
SiO_2	0.64%	2.13%	1.31%	5.63%
Al_2O_3	0.19	0.18	0.34	0.19
Fe_2O_3	0.59	0.46	0.65	0.48
TiO_2	0.00	0.01	0.01	0.02
ZrO_2	--	--	--	11.39
CaO	0.89	1.35	1.68	0.40
MgO	97.38	95.66	95.79	81.70
Na_2O	0.10	0.02	0.06	0.07
K_2O	0.00	0.00	0.00	0.00
Total:	99.79%	99.81%	99.84%	99.87%
CaO/SiO_2	1.39	0.63	1.28	0.11

Table 4. Linear change (%) after simulative checker testing				
	High Purity Burned Magnesia Brick	Intermediate-Purity Burned Magnesia Brick #1	Intermediate-Purity Burned Magnesia Brick #2	Magnesia-Zircon Brick
% Linear Change				
Cycled	-0.55	-0.06	-0.27	+0.16
Cycled with Sulfate	+0.39	+0.58	-0.63	+0.03

Table 5. Weight change (%) after simulative checker testing				
	High Purity Burned Magnesia Brick	Intermediate-Purity Burned Magnesia Brick #1	Intermediate-Purity Burned Magnesia Brick #2	Magnesia-Zircon Brick
% Weight Change				
Cycled	+0.78	+2.98	+2.28	+0.15
Cycled with Sulfate	+1.07	+2.47	+6.52	+1.24

The as-received brick and the test samples were analyzed for Na_2O content to try to explain the weight gains during the test. As shown in Table 6, the highest Na_2O contents were for the brick cycled in contact with sodium sulfate. The samples tested in the "closed" container that contained no sodium sulfate also exhibited increases in Na_2O content. This seemed to imply that the container in which those samples were held was not completely sealed, and that the samples therefore had some exposure to sodium sulfate that was present in the atmosphere of the furnace.

Table 6. Weight % Na$_2$O in as-received and test samples			
Weight % Na$_2$O	Unused	Cycled	Cycled with Sodium Sulfate
High Purity Burned Magnesia Brick	0.10	0.20	0.98
Intermediate-Purity Burned Magnesia Brick #1	0.02	0.81	1.90
Intermediate-Purity Burned Magnesia Brick #2	0.06	0.43	3.03
Magnesia-Zircon Brick	0.05	0.07	1.43

Table 7 gives a comparison of modulus of rupture at 1038°C for the samples after the simulative checker testing. Compared to the samples that were cycled with no contact with sodium sulfate, there were strength losses in all cases. The magnesia-zircon brick excelled in strength retention when in contact with sodium sulfate.

Table 7. Modulus of rupture at 1038°C after simulative checker testing				
	High Purity Burned Magnesia Brick	Intermediate-Purity Burned Magnesia Brick #1	Intermediate-Purity Burned Magnesia Brick #2	Magnesia-Zircon Brick
MOR at 1038°C, N/mm^2				
Cycled	16.4	5.93	7.38	17.2
Cycled with Sulfate	2.76	2.14	5.03	11.3

Chemical analyses were run on the unused brick and on samples after testing. As shown in Table 8, there were substantial reductions in CaO content of the burned magnesia brick that had been in contact with sodium sulfate during cycling. There was not a reduction in CaO content for samples that were not in contact with sodium sulfate. A small reduction in CaO was measured for the magnesia-zircon brick tested in contact with sodium sulfate. Still, because of the high amount of SiO$_2$ in the magnesia-zircon brick, there was not a substantial change in lime-to-silica ratio of that product during the test. The loss of CaO from the burned magnesia brick caused substantial changes in the lime-to-silica ratios, which would be expected to alter the composition of the bond phases of those brick.

Table 8. CaO contents after simulative checker testing			
	Unused	Cycled	Cycled with Sodium Sulfate
High Purity Burned Magnesia Brick			
% CaO	0.89	0.99	0.38
CaO/SiO$_2$ Ratio	1.39	1.45	0.56
Intermediate-Purity Burned Magnesia Brick #1			
% CaO	1.35	1.44	0.38
CaO/SiO$_2$ Ratio	0.63	0.68	0.17
Intermediate-Purity Burned Magnesia Brick #2			
% CaO	1.68	2.14	0.97
CaO/SiO$_2$ Ratio	1.28	1.51	0.70
Magnesia-Zircon Brick			
% CaO	0.40	0.40	0.30
CaO/SiO$_2$ Ratio	0.11	0.07	0.04

The x-ray diffraction results shown in Table 9 showed that all samples that had been in contact with sodium sulfate contained the mineral thenardite (Na_2SO_4). This was consistent with the Na_2O analyses reported in Table 6, and indicates that heating the furnace to 1177°C after cycling the samples did not result in volatilization of all the sodium sulfate. Loss of CaO from the burned magnesia brick pushed the mineralogy of the bond phase in the direction of forsterite ($2MgO \cdot SiO_2$), as would be expected. In the magnesia-zircon brick, the only silicate phase detected was forsterite, and unlike the case with the burned magnesia brick, exposure to sodium sulfate did not result in detection of new bond phases. Interestingly, the unused magnesia-zircon brick contained cubic zirconia, while the samples run in the simulative checker test contained a combination of cubic and monoclinic zirconia. There is a possibility that CaO contributed to stabilization of the cubic zirconia in the unused brick, and that migration of CaO during testing caused destabilization of a portion of the cubic zirconia and appearance of monoclinic zirconia.

In addition, the absence of formation of magnesium sulfate, $MgSO_4$, in all the samples is noteworthy, as formation of magnesium sulfate has been cited as a wear mechanism [9].

The results of the simulative checker testing support the current practice of using magnesia-zircon brick in the condensate zone of glass tank checkers. The greater strength retention of the magnesia-zircon brick after cycling compared to the burned magnesia brick, and the depletion of CaO from magnesia brick in the presence of sodium sulfate favor the magnesia-zircon brick.

Table 9. X-ray diffraction results			
	Unused	Cycled	Cycled with Sodium Sulfate
High Purity Burned Magnesia Brick			
CaO/SiO$_2$ Ratio:	1.39	1.45	0.56
	MgO	MgO	MgO
	C$_3$MS$_2$	C$_3$MS$_2$	CMS
	--	Possible M$_2$S	M$_2$S
	--	--	Na$_2$SO4
Intermediate-Purity Burned Magnesia Brick #1			
CaO/SiO$_2$ Ratio:	0.63	0.68	0.17
	MgO	MgO	MgO
	CMS	CMS	CMS
	M$_2$S	M$_2$S	M$_2$S
	--	--	Na$_2$SO$_4$
Intermediate-Purity Burned Magnesia Brick #2			
CaO/SiO$_2$ Ratio:	1.28	1.51	0.70
	MgO	MgO	MgO
	CMS	CMS	CMS
	C$_3$MS$_2$	Possible M$_2$S	Na$_2$SO4
Magnesia-Zircon Brick			
CaO/SiO$_2$ Ratio	0.11	0.07	0.04
	MgO	MgO	MgO
	CZ	CZ	CZ
	M$_2$S	MC	MZ
	--	M$_2$S	M$_2$S
MgO = periclase	--	--	Na$_2$SO$_4$
C$_3$MS$_2$ = merwinite			
CMS = monticellite			
M$_2$S = forsterite			
CZ = cubic zirconia			
MC = monoclinic zirconia			
Na$_2$SO$_4$ = thenardite			

SIMILARITY TO BASIC REFRACTORIES USED IN THE CEMENT INDUSTRY

Depletion of CaO from magnesia brick in the presence of sodium sulfate was observed with the aforementioned used checker brick and was simulated in laboratory testing by cycling burned magnesia brick in contact with sodium sulfate. Often to support observations or to verify the existence of certain phenomena refractory engineers will look to applications with similar conditions. There is evidence from the cement industry that CaO depletion can occur in magnesia-spinel brick in a sulfate-rich environment. Cement kilns often burn petroleum coke and used tires, which contain significant sulfur. In addition, fly ash from power plants is often a constituent of the cement formulation. Fly ash contains significant K$_2$O. As a result of these factors, basic brick in rotary cement kilns are exposed to significant SO$_3$ and K$_2$O.

Figure 3 shows the cross-section of a magnesia-spinel brick after use in a rotary cement kiln. The photograph shows the worn hot face and the location of various samples which were taken for chemical analysis.

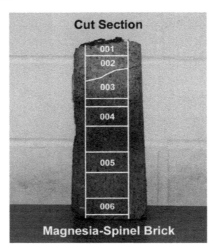

Figure 3. Magnesia-spinel brick after service in a rotary cement kiln

Table 10 shows the chemical analyses of the magnesia-spinel brick samples taken from various locations in the used brick compared to the chemistry of a typical unused brick of the same composition. All oxide analyses were done on a calcined basis except for SO_3, Na_2O and K_2O which were done on an as-received basis because those constituents are volatile and could be lost when calcining the samples. Near the cold face of the brick the SO_3 content rose substantially, as did the K_2O content. The Na_2O content did not increase greatly at the cold face of the brick. These results suggested condensation of K_2SO_4 rather than Na_2SO_4. Examination of the CaO contents shows depletion of CaO at the hot face of the brick, and an increase in CaO at the location near the cold face where K_2O and SO_3 reached the highest levels. These results seemed to parallel the situation seen in the aforementioned burned magnesia brick taken from the glass tank regenerator where it was in contact with sodium sulfate.

Table 10. Chemical analysis of magnesia-spinel brick after service in a rotary cement kiln							
Sample Designation:	1	2	3	4	5	6	Unused
Location	Hot Face	Interior	Interior	Interior	Interior	Cold Face	
Chemical Analysis (Calcined Basis)							
SiO_2	0.25%	0.28%	0.19%	0.21%	0.20%	0.31%	0.23%
Al_2O_3	5.84	6.76	6.46	6.36	6.44	6.35	7.17
Fe_2O_3	0.54	0.47	0.40	0.47	0.46	0.45	0.21
CaO	0.20	0.15	0.15	0.16	1.87	1.00	1.16
MgO	92.7	91.5	91.8	89.0	86.6	91.1	91.23
MnO	0.05	0.03	0.04	0.06	0.07	0.07	--
Chemical Analysis (As-Received Basis)							
SO_3	0.55	0.67	0.13	0.06	4.99	2.56	--
Na_2O	0	0	0	0.02	0.02	0.02	--
K_2O	0.51	1.02	1.47	6.05	7.33	0.96	--

ATTACK OF MAGNESIA-ZIRCON BRICK BY Na_2O

While forsterite-bonded magnesia-zircon brick are appropriate for condensation zones in regenerators operated under oxidizing conditions, it has been reported that under reducing conditions magnesia-zircon brick are attacked by NaOH in the waste gas stream [9]. The theory is that exhaust gases from the glass furnace contain SO_2 from the sodium sulfate and/or from the fuel. Under oxidizing conditions, there is enough oxygen in the exhaust gases to convert the SO_2 to SO_3, and soda in the exhaust gases combines with the SO_3; this results in deposition of Na_2SO_4 in the condensate zone of the regenerator. Under reducing conditions, however, the SO_2 cannot oxidize to SO_3, preventing it from combining with the soda to form Na_2SO_4. This creates the possibility of having free soda in the exhaust gas stream that can react with magnesia-zircon brick in the condensate zone.

Another environment where excess Na_2O might be expected to exist in the exhaust gases is in sodium silicate furnaces. Sodium silicate furnaces contain a substantial amount of sodium carbonate, Na_2CO_3, as a batch constituent. The large amounts of sodium carbonate in sodium silicate batches raises concern about reaction of excess Na_2O in the exhaust gases with the refractories, similar to the situation in flat glass and container glass furnaces operated under reducing conditions.

A magnesia-zircon brick had been installed as the rider arches, rider tile and transition tile in the regenerator of a sodium silicate furnace. After 3 years of service, the magnesia-zircon brick courses were found to have undergone severe reaction, causing cracking and spalling of the arches and tile. As illustrated in Table 11, a used magnesia-zircon brick from the furnace had a Na_2O content of 30.0%, indicating a high amount of pickup of that constituent. The SO_3 level was only 2.4%, indicating that the Na_2O level was far too high to be attributable to condensation of Na_2SO_4. The high amount of Na_2O pickup seemed to parallel the situation that could occur under reducing conditions in container glass and flat glass furnaces.

Table 11. Chemistries of used magnesia-zircon brick after service in a sodium silicate furnace regenerator compared to data sheet properties

Sample:	Used Magnesia-Zircon Brick	Comparative Data Magnesia-Zircon Brick Data Sheet Properties
Chemical Analysis*	(semi-quanitative)	
(As Received Basis)		
Silica (SiO_2)	5.7%	8.2%
Alumina (Al_2O_3)	0.33	0.2
Iron Oxide (Fe_2O_3)	0.54	0.1
Zirconia (ZrO_2)	16.7	12.3
Lime (CaO)	0.47	0.5
Magnesia (MgO)	45.9	78.7
Soda (Na_2O)	30.0	---
Potash (K_2O)	0.02	---
Total Analyzed	99.66%	100.0
Sulfur Trioxide (SO_3),%	2.4	---

X-ray diffraction was run to detect the phases in the used magnesia-zircon brick from the checker setting of the sodium silicate furnace. The results are presented in Table 12.

Table 12. X-ray diffraction results of a magnesia-zircon brick after service in a sodium silicate furnace regenerator

	Used Magnesia-Zircon Brick
Periclase (MgO)	Major
Thenardite (Na_2SO_4)	Trace?
Anhydrite ($CaSO_4$)	Trace?
Albite ($NaAlSi_3O_8$)	Trace?
Baddeleyite (ZrO_2)	Trace-minor
Bredigite ($Ca_{14}Mg_2[SiO_4]_8$)	Trace?
Calcium Zirconate ($CaO \cdot ZrO_2$)	Trace
$Ca_{0.15}Zr_{0.85}O_{1.85}$	Trace
Merwinite ($3CaO \cdot MgO \cdot 2SiO_2$)	Trace
Sodium zirconium silicate (Na_2ZrSiO_5)	Minor
Parakeldyshite ($Na_2ZrSi_2O_7$)	Trace

Because periclase (MgO) was the dominant phase in each used brick, the peaks of the other phases were small, making it more difficult to identify the secondary phases.

An unused magnesia-zircon brick would contain substantial amounts of periclase (MgO), forsterite ($2MgO \cdot SiO_2$) and zirconia (ZrO_2). The x-ray diffraction results on the used magnesia-zircon brick revealed no forsterite and trace to minor amounts of baddeleyite (ZrO_2). However, two sodium-zirconium-silicate phases were detected, sodium zirconium silicate (Na_2ZrSiO_5) and parakeldyshite ($Na_2ZrSi_2O_7$). The high level of Na_2O (30%) in the used magnesia-zircon brick along with the absence of forsterite suggested alteration of the forsterite into a glassy alkali-containing phase. The presence of the two sodium-zirconium-silicate phases was an indication that the zirconia as well as the forsterite entered into a reaction with the Na_2O that infiltrated the brick.

To better understand the effect of the zirconia portion of a magnesia-zircon brick on alkali resistance, a test was run exposing brick samples to sodium carbonate. The following brick compositions were tested: (1) a magnesia-zircon brick, (2) a magnesia-forsterite brick that did not contain ZrO_2 and (3) a burned magnesia brick. Twenty-five grams of sodium carbonate were added to an alumina crucible and brick samples measuring 62 x 62 x 25 mm were placed on top of the crucibles with magnesia spacers to prevent reaction of the brick with the crucible. The samples were heated at 90°C/hour to 1427°C, where they were held for 48 hours. The idea was that the sodium carbonate would dissociate in heating, and that the test samples would be exposed to Na_2O in the vapor phase.

As shown in Table 13, the magnesia-zircon brick exhibited 6.7% volume expansion during the test, while the magnesia-forsterite brick without zirconia exhibited only 0.93% volume expansion. The volume expansion of the magnesia-forsterite brick without zirconia approached the 0% volume expansion exhibited by the burned magnesia brick sample, which was known to have good alkali resistance.

Table 13. Alkali test results			
	Magnesia-Zircon Brick	Magnesia-Forsterite Brick Without ZrO_2	Burned Magnesia Brick
Alkali Vapor Test (Av 2)			
Volume Change, %:	6.70	0.93	0
Chemical Analysis Before Testing (Calcined Basis)			
SiO_2	6.87%	6.18%	0.91%
Al_2O_3	0.25	0.24	0.19
TiO_2	0.05	0.01	0.00
Fe_2O_3	0.16	0.25	0.31
ZrO_2	13.12	0.00	0.05
CaO	0.85	1.19	2.27
MgO	78.63	92.12	96.04
MnO	0.01	0.00	0.12
Na_2O	0.00	0.00	0.00
K_2O	0.00	0.00	0.00
P_2O_5	0.04	0.00	0.00
Total	99.99%	99.99%	99.99%

In an attempt to determine why removal of the zirconia component so dramatically reduced the volume expansion of a magnesia-forsterite brick in the alkali vapor test, a detailed mineralogical examination was conducted. The altered portion of the magnesia-zircon brick showed the presence of a complex Na-Mg-Al-Si-Zr-Ca-O phase or phases in the matrix of the brick after testing. There was significant opening of the matrix at the surface of the magnesia-zircon brick in contact with the alkali vapors (See Figure 2). The forsterite-bonded brick that contained no ZrO_2 had a much denser hot face as shown in Figure 4, although some cracking was evident at high magnification as shown in Figure 5. The fine cracks may have been associated with the expansion of the magnesia-forsterite brick in the alkali vapor test, but the much smaller degree of matrix opening in the magnesia-forsterite brick likely accounted for the substantially lower expansion of that brick compared to the magnesia-zircon brick. The presence of Zr in the altered matrix phases of the magnesia-zircon brick supported the conclusion that ZrO_2 reacts with alkali. The lack of ZrO_2 in the magnesia-forsterite brick apparently reduced the amount of constituents that were potentially reactive with and alterable by the alkali. The presence of Al in the altered samples suggested either that some of the Al_2O_3 from the crucibles in the test transferred to the samples, or that there was residual Al_2O_3 present from the medium used to polish the samples prior to microscopic examination. Figures 6-8 show microstructures of forsterite-bonded magnesia brick after alkali vapor test.

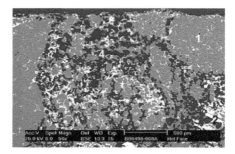

Figure 4. Magnesia-zircon brick after alkali vapor test (altered hot face) 50X

1. Remnant coarse magnesia grain
2. Zirconia
3. Matrix bonded by complex Na-Mg-Al-Si-Zr-Ca-O Phase(s)
4. Void

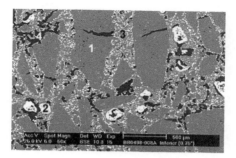

Figure 5. Magnesia-zircon brick after alkali vapor test (unaltered interior)

1. Coarse magnesia grain
2. Zirconia
3. Forsterite bonded matrix
4. Void

Figure 6. Forsterite-bonded magnesia brick with no ZrO_2 after alkali vapor test

(Altered hot face) 50X

1. Remnant coarse magnesia grain
2. Matrix bonded by Na-Mg-Al-Si-Ca-O phase(s)
3. Void

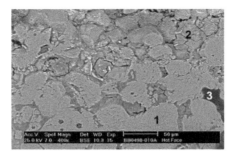

Figure 7. Forsterite-bonded magnesia brick with no ZrO$_2$ after alkali vapor test showing fine cracks

(Altered hot face) 400x

1. Periclase crystallite
2. Na-Mg-Al-Si-Ca-O phase(s)
3. Void

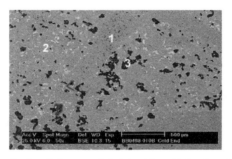

Figure 8. Forsterite-bonded magnesia brick with no ZrO$_2$ after alkali vapor test

(Unaltered cold face) 50x

1. Coarse magnesia grain
2. Forsterite bonded matrix
3. Void

These results suggested that a forsterite-bonded brick containing no ZrO$_2$ might be a viable substitute for a magnesia-zircon brick in condensate zones, particularly in gas-fired furnaces. It is uncertain whether a forsterite-bonded brick with no ZrO$_2$ would be appropriate for flat glass and container furnaces that were intentionally run under reducing conditions, since it would still be possible for alkali to attack the forsterite phase. Nevertheless, the ZrO$_2$-free brick might be considered to be ideal for regenerators run under oxidizing conditions.

Alkali resistance, of course, is not the only property of importance for a forsterite-bonded brick. Physical properties, including creep resistance, are of high interest. As shown in Table 14, the properties of the magnesia-forsterite brick without ZrO_2 were comparable to those of the magnesia-zircon brick. Modulus of rupture at 1482°C was somewhat lower for the magnesia-forsterite brick without zirconia, but the strength was considered to be adequate for a condensate zone.

Table 14. Properties of the bricks		
	Magnesia-Zircon Brick	Magnesia-Forsterite Brick Without ZrO_2
Data from Porosity Test (Av 5)		
Bulk Density, g/cm³:	3.09	2.93
Apparent Porosity, %:	14.5	15.5
Apparent Specific Gravity:	3.61	3.47
Modulus of Rupture, N/mm² (Av 5):		
At Room Temperature:	14.7	15.2
At 1260°F:	16.1	10.9
At 1482°F:	10.5	5.2
DIN Creep Test at 1500°C Using 0.2 N/mm² load		
Dmax,% [10]:	1.56	1.87
Tmax, °C [11]:	1308	1403
% Subsidence in 0 to 50 Hours:	4.25	3.67
% Subsidence in 5 to 25 Hours:	1.71	1.50
% Subsidence in 20 to 50 Hours:	1.22	1.24
Cold to Cold Dimensional Change:	-4.02	-3.52
Chemical Analysis (Calcined Basis)		
Silica (SiO_2)	6.32%	5.18%
Alumina (Al_2O_3)	0.42	0.71
Titania (TiO_2)	0.03	0.01
Iron Oxide (Fe_2O_3)	0.17	0.30
Zirconia (ZrO_2)	12.16	0.0
Lime (CaO)	0.79	0.98
Magnesia (MgO)	80.04	92.73
Manganese Oxide (MnO)	0.02	0.02
Soda (Na_2O)	0.05	0.11
Potash (K_2O)	0.00	0.0
Phosphorous Pentoxide (P_2O_5)	0.04	0.0
Total	100.03%	100.04%

CONCLUSIONS

This paper described some of the principal wear mechanisms of basic refractories used in glass tank regenerators. Alteration of refractories by V_2O_5 and depletion of CaO from refractories exposed to sodium sulfate was discussed. Alkali attack on forsterite-bonded basic brick was described. Properties of a new type of forsterite-bonded brick that offers the potential for improved performance were presented.

REFERENCES

1. M. Eke and N. H. Brett, "Phase Equilibria in the System $CaO-MgO-SiO_2-V_2O_5$: Their Relevance to Refractory Usage in Glass Tank Regenerators - I. The Particle Systems $MgO-C_2S-V_2Os$ and $MgO-C_3S-V_2Os$ at 1500°C," *Trans. Brit. Ceram. Soc.* **72**(5), 195 (1973).

2. E. F. Osborn, "The Compound Merwinite ($3CaO \cdot MgO \cdot 2SiO_2$) and Its Stability Relations Within the System $CaO-MgO-SiO_2$ (Preliminary Report)," *J. Am. Ceram. Soc.*, **26**(10), 321-332 (1943).

3. R. A. McCauley, "The Effects of Vanadium Upon Basic Refractories," in *Proceedings of UNITECR*, 858-865 (1989).

4. M. Eke and N. H. Bret, op. cit.

5. M. Carter and N. H. Brett, "Phase Equilibria in the System $CaO-MgO-SiO_2-V_2O_5$: Their Relevance to Refractory Usage in Glass Tank Regenerators, II. The Partial Systems $MgO-M_2S-V_2O_5$, $MgO-C_2S-V_2O_5$, $MgO-C_2S-C_3V$ and $MgO-50\% C_2S$, $50\%C_3S-V_2O5$ at 1450°C," *Trans. Brit. Ceram. Soc.*, **72**(5), 203-207 (1973).

6. Dicalcium silicate is represented as Ca_2SiO_4 in Figure 1.

7. Forsterite is represented as Mg_2SiO_4 in Figure 2.

8. J. R. LeBlanc, "Brockway's Lower Checker Sulfate Test," *Journal of the Canadian Ceramic Society*, **52** (1983).

9. G. Heilmann et. al, "New Solutions for Checkers Working Under Oxidizing and Reducing Conditions," in *Proceedings of the 67th Conference on Glass Problems*, 183-195 (2007).

10. D_{max} is the maximum % increase in sample height due to thermal expansion during the test.

11. T_{max} is the temperature at which maximum expansion of the sample occurs.

MODERN AND COMPETITIVE REGENERATOR DESIGNS FOR GLASS INDUSTRY

Sébastien Bourdonnais
Saint-Gobain SEFPRO
Le Pontet, France

ABSTRACT

This paper presents an overview of generators with design recommendations. Checker types and common solutions will be summarized. In addition, materials choice and lifetime resistance will be discussed.

INTRODUCTION

Modern soda lime glass manufacturing has become a challenging global market. Energy consumption and environmental regulation are now major concerns. To improve the competitiveness, operating a high performance trouble-free regenerator is a fundamental point. In addition to obvious requirements of regenerative efficiency, it is observed globally that more and more campaigns are interrupted after corrosion issues with regenerators. Indeed, number of different destructive phenomena that can occur in these regenerators is large and evolve with new glass practices such as typical raw material choice or combustion settings.

This paper gives some general considerations about the state-of-the-art practices with design advices and choices of materials based on worldwide furnace park observation as well as internal expertise. The purpose is to help glassmakers to make proper choices when a repair or a green field construction is foreseen. After a short general overview about regenerators, the first part of this paper will deal with design recommendations for regenerators and describe the important criteria to reach satisfying efficiency. Then the choice of checkers' types will be documented with the description of the common solutions available on the market. Finally, the material choice for both chambers' walls and checkers will be discussed, based on the lifetime resistance priority.

IMPORTANCE OF A REGENERATIVE SYSTEM

It is not useless to remind to non-experts the fundamental necessity of having a regenerative system embedded with the furnace. One could wonder why regenerators have been quickly added to the core furnace at the early age of the modern glass industry. Obviously, such an impressive structure in addition to the melting and the forming areas has many consequences on the investment and facilities requirements. Moreover, the numerous technical concerns related to regenerator exploitation might make glassmakers tempted to manage without them. Figure 1 shows that a furnace without any system of air preheating would see its consumption increase drastically.

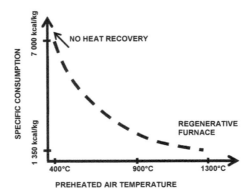

Figure 1. Consumption versus preheating air temperature

The calories recovered from the hot gases and used to heat the combustion air allow a reduction of the supply of fossil fuel by 5 to even 10 times depending on the case. Due to the cost of common energy, the payback of the regenerative system is usually reached in the first months of exploitation, whatever the configuration is. If we exclude from this discussion the case of oxy-fuel, which is attractive in a few world regions only, the regenerative furnace remains today the first technology to melt soda lime glass at an industrial scale.

Figure 2 describes the global thermal balance of a regenerative furnace to define the criteria of efficiency. Even though a side-port furnace is shown, the principles hereafter are valid for end-port furnaces too.

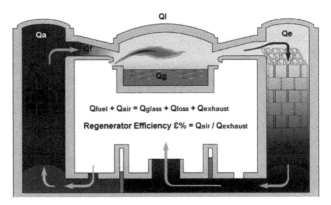

Figure 2. Furnace thermal balance overview

Considering the following equilibrium between the heats supplied into the laboratory (Q_{fuel} and Q_{air}) and the heats used after combustion (Q_{glass} for melting the glass, $Q_{exhaust}$ released in the fumes and Q_{loss} representing the various thermal losses), we call Regenerator Efficiency the ratio between Q_{air} and $Q_{exhaust}$. The higher this ratio is, the more the heat from the combustion fumes is recovered. The theoretical maximum based on thermodynamics is around 80%. The common industrial regenerators usually operate between 55 and 75%.

GENERAL PRINCIPLE FOR CHAMBER DIMENSIONING

For a green field project, the initial step of the regenerator design phase usually starts with the chamber volume definition. Some relations between the pull and the total chamber volume for both sides exist. Twenty years ago, the typical relation was Volume = Pull/0.6. More recent furnaces operate today with much bigger chambers than the ones designed in the 90's. Table 1 below illustrates the example of float furnaces, comparing the regenerator size and the specific consumptions associated.

Table 1. Float furnaces, 90's design versus 2010's

	PULL (TPD)	CHAMBER VOLUME BOTH SIDES (M3)	CHECKER HEIGHT (M)	SPECIFIC CONSUMPTION (BTU/ GLASS US TONS)
Design 90's #1	520	950	7,6	5 580
Design 90's #2	550	1 100	7,3	5 580
Improvements :				
2008 Europe #1	600	1 600	9	4 680
2009 Europe #2	600	1 350	11	4 680
2011 Europe #3	570	1 450	11	4 750

In the last 20 years, for a similar range of daily pull, the size of regenerators has increased by 50% on average while specific consumptions have decreased by 15%. The 90s' drastic increase of energy price and the progress in terms of performance prediction systems have led to some paradigm shifts in the chamber dimensioning. Recent float furnaces in Europe have much bigger chambers with checkerpacks higher than 9 meters, as a consequence, these modern furnaces often reach historically low specific consumptions, below 4 700 BTU/ Glass US Tons (1 300kcal/glass kg).

Once the overall chamber volume is estimated, the main discussion is related to the 3 dimensions definition. Several parameters should be taken into account at this stage. The available space is obviously one of them, particularly when repairing an existing furnace in a given plant. In the case of end-port furnaces, one constraint is also the width of the furnace itself, it is not recommended to have chambers larger than the furnace. On top of a complex layout, one problem in that case would be the homogeneity of the exhaust fumes' repartition, this chamber's extra-width would certainly lead to a dead zone with a limited amount of preheating.

In case of new constructions or of limited space constraint, when starting from a blank page in the regenerator's design, a good way to proceed is to first size the section of the chamber so as to reach optimal velocities of air and exhaust fumes. The velocity is calculated by dividing the fluid flow (air or fumes) with the open section of the regenerator (sum of open flues). It has been demonstrated that a very high velocity of fluid usually leads to a poor overall efficiency. Basically, the checker pack needs enough time to capture the energy from the hot gases and to preheat the air properly. For instance, an optimal velocity for exhaust gases is usually in the range of 0.25-0.35 Nm/s. Below this range, with "oversized regenerator" the efficiency will be good anyway but the extra investment (for refractories and civil engineering) may not have a short payback. Above this optimal range of values, the regenerator will be considered undersized and the efficiency might be low.

Another critical point is related to the depth by width ratio of the section. Too deep a chamber could generate a preferential path for fumes on the target wall side, leaving a much colder area below the port neck. This heterogeneous preheating of the packing would, of course, badly affect the temperature of the combustion air. In the case of an end-port furnace, a 1.3-1.4 ratio between Depth and Width is commonly recommended. Some industrial furnaces with a ratio higher than 1.7 are considered as affected by the aforementioned problem.

The previous remarks related to an optimized section definition are of course valid for a usual given height. If compatible with the available free room, the increase in height is, of course, a good way of enhancing overall efficiency. Figure 3 below illustrates a theoretical example of regenerator efficiency improvement as a function of the increase in the checker pack's height.

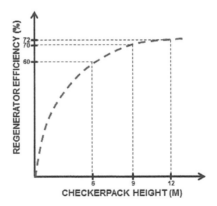

Figure 3. Example of checker pack height versus regenerator efficiency
(Constant section)

The relation between height and efficiency is asymptotic, meaning that incremental steps of height will not proportionally improve efficiency. At some point, when approaching the theoretical maximum, the improvement is insignificant versus the extra charge of investment. Generally speaking, modeling is a good tool to simulate the various scenarios and helping decide which height for a checker pack will represent the best cost/performance ratio. Nowadays, the majority of the modern soda lime furnaces have checker packs measuring up to between 8 and 11 meters.

CHOICE OF CHECKERS' DESIGN

In a given chamber, the choice of checkers' type is a fundamental parameter that will drive both the regenerator's efficiency and its lifetime. This part of the paper will only discuss the various shapes of common solutions available on the market and their impact on regenerators' efficiency. In fact, it is commonly admitted that the geometric properties of the checkers mainly impact the regenerator's efficiency while the chemical nature of the checkers gives the corrosion resistance to the packing, that is to say its lifetime.

Figure 4 below presents a selection of common checkers' types available on the market from different manufacturers of refractories. Considerations of chemistry or manufacturing process aside, these products differ by the shape of their unit. From very simple bricks to a textured cross-shaped piece, the geometry of these products really impacts their contribution in the thermal exchanges.

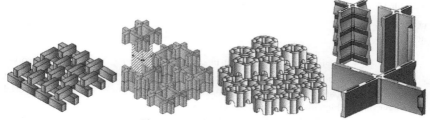

Figure 4. Various common checkers designs

From Left to Right: 'Maerz' bricks, Interlock "H" bricks, Chimney blocks, and Cruciforms

A very simplified expression of the Hausen law for thermal exchanger, which describes the thermal exchanges, is given below:.

$$Q = K.S.\Delta\theta$$

This relation defines the quantity of heat exchanged per regenerative cycle (Q) as being a function of 2 parameters. The first parameter is the Specific Surface (S), this property quantifies the surface of checkers per chamber's volume unit (for instance Ft^2/Ft^3). Basically, the higher this S value is, the more efficient the heat storage/restitution will be. The Specific Surface is calculated knowing both unit brick's surface but also bricks numbers in the packing. In fact, the checkers' pattern will impact the number of flues available and then, the overall specific surface. Some checkers have flue size adaptability so datasheets need to be compared in details when benchmarking checkers' solutions. Table 2 below gives some typical values for the products mentioned above.

Table 2. Various common checkers geometrical properties

PRODUCT NAME	TYPICAL NOMINAL FLUE SIZE (MM)	CHECKER THICKNESS (MM)	SPECIFIC SURFACE (M^2/M^3)
Basket Weave Bricks	150 x 150	64	11.0
Straight Pigeon Hole Bricks	140 x 140	64	12.7
Maerz Diagonal Staggered Bricks	150 x 150	76	14.0
Interlock "H" bricks	146 x 146	44.5	17.4
Chimney Blocks TG	140 x 140	40	15.9
Chimney Blocks TL	140 x 140	40	16.9
Cruciform T4	140 x 140	30	20.0
Cruciform T4	150 x 150	30	18.1
Cruciform T8	150 x 60	30	26.0
Cruciform T6	330 x 330	30	9.3

Even if such design is still widely used in some world regions, every checker works' concepts based on straight bricks has rather low Specific Surface. The Maerz design (diagonal staggered pigeon hole pattern) is the best way to assemble straight bricks to increase exposed area. However, both Chimney Blocks and Cruciforms solutions have better properties and are today considered as standard solutions for modern furnaces. Straight bricks are penalized by their higher thickness that reduces the open flue area in a given section. The highest Specific Surface is obtained with Cruciforms' Types combination.

Selecting checkers with the flue size flexibility allows a better overall optimization of the system. Indeed, the importance of having a higher specific surface at the top of the packing rather than at the bottom has been demonstrated in the past. This can be explained by the various flow regimes within the height of the regenerator. The natural convection at the bottom does not require so much exchange surface than the forced convection above.

The second parameter from the Hausen law to be taken into consideration is called the global heat exchange coefficient (K). This parameter is as much contributive to the overall heat exchange as the specific surface despite being less easy to quantify. This global coefficient K is indeed a combination of several parameters such as thermal conductivity of the checkers, thickness of the bricks and mainly heat exchange coefficient between fumes and refractories (during fumes cycle) and heat exchange coefficient between refractories and air (during air cycle).

These heat exchange coefficients describe the quality of the physical contact between the fluid and the checkers. Beyond the value of the specific surface itself, what matters is "how" the surface is in contact with the fluids (fumes then air). For instance, some textured checkers, drastically increase the turbulences at their interface with the air to be preheated, increasing the quality of the contact, so the heat exchange coefficient.

Figure 5 illustrates the effect of an obstacle at the surface of the checker, in a laminar flow for a given channel. Locally, the generated turbulence is drastically increasing the heat transfer coefficient (h) between the 2 elements (refractory checker and air) compared to a smooth checker case ($h0$).

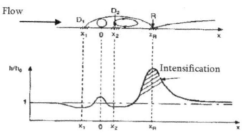

Figure 5. Intensification of heat transfer when the fluid locally hits an obstacle

Textured checkers like Cruciforms Type 4 (Figure 6) have corrugated wings that generate turbulences inside the flue.

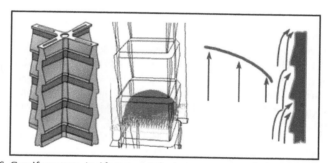

Figure 6. Cruciform type 4 with corrugated wings generates turbulences within the flue

When placed at the middle and the top part of the checker pack, such checkers improved the heat transfer quality within a forced convection regime. The improvement of the regenerator's efficiency with such checkers compared with smooth checkers like bricks or chimney blocks can reach several percents, as reported in the past.

One last criterion in terms of the choice of checker's design is related to the flow repartition within the section of the chamber. It is fundamental to achieve a homogeneous circulation of hot fumes within the checker pack's section to get an optimal air preheating in the end. As previously mentioned, an appropriate chamber's Depth/Width ratio is one way of reducing the natural heterogeneity of exhaust gases' flow from target wall to port neck. The use of checkers that form an open flues pattern can then be considered. Figure 7 shows a top view comparison between with open flues which allow a horizontal flow movement and chimney blocks with closed flues. With Cruciforms, the natural fumes flow gradient that occurs at the very top of the checker pack is absorbed within the first top courses. When reaching the equivalent of 1 meter of height, the balance of the flow's repartition is drastically improved with open flues.

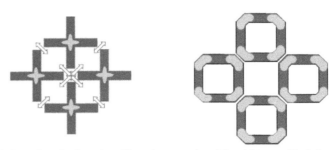

Figure 7. Open flue checkers (cruciforms) versus closed flue (chimney blocks)

CHOICE OF MATERIALS FOR CHAMBER AND CHECKERS

When the chamber's dimensioning and the checkers' design choice are done, then comes the refractory's choice selection. Of course, to properly select the materials among the large choice of references available on the market, a quick overview of the typical corrosion mechanisms in the chambers is useful. Figure 8 is used below to describe the commonly admitted reactions of corrosion.

The various mechanisms will be commented from top to bottom, based on the areas of temperatures. Some suggestions about the possible choice of the refractories will be discussed in relation with the given comments.

The very top part, above 1 350°C, is the area where both chambers (crown and upper part of the walls) and checkers (first top meter) are exposed to the harshest environment. In this area, the exhaust gases enter the chamber at the highest temperature, fully loaded with the carry-over of raw materials. Possible corrosion mechanisms are numerous, leading to excessive wear, checkers plugging or damaging of walls including by-pass between the two sides of an end-port regenerator.

The various batch compositions may lead to extremely different situations and somehow quite unpredictable ones. At these high temperatures, typical aggressive raw materials like dolomite or lime represent the typical threat. Glass cullet's size and ratio as well as batch moisture are the most noticeable recent evolutions in the batch composition. More and more very fine glass cullet's quantities, combined with a dry batch may generate extremely aggressive carry-over for conventional refractories. The key problem in this area of the regenerator is the difficulty to find one single refractory, which resists to the many various aggressive agents. For instance, Fused Cast AZS is very good to resist glass cullet's corrosion but remains pretty weak against soda vapors, whose concentration is high in the fumes. The use of heavy oil of petcoke's combustibles also impacts the corrosiveness of the fumes since their vanadium content is quite dreadful.

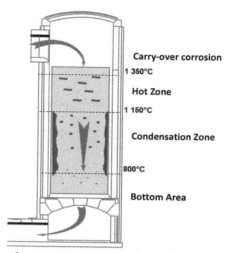

Figure 8. Soda lime furnace regenerator, corrosion mechanisms areas

This area is the most sensitive part of the regenerator when it comes to refractory's choice, and quite often some technical compromises are somehow required. Among the several options that coexist on the market today, the use of high-quality mullite or sillimanite-mullite for walls and crown represents a good balance between the various criteria required (resistance to corrosion and mechanical properties). Silica walls and crown offer pretty low investment cost but really weak resistance to corrosion which makes expensive hot repairs often mandatory to finish the campaign. Basic refractory walls with high-magnesia products can offer variable performances; numerous cases of creeping resistance issue were reported. Sometimes, only the target wall (fumes' impact area) is reinforced with most noble refractories. For the checkers, fused cast alumina based checkers are a very good versatile candidate to properly match the various situations described above, magnesia-zirconia checkers are also a common solution with acceptable performances but disadvantaged by their bonded nature (high open porosity content).

The area between 1350°C and 1150°C is usually less critical in terms of corrosion. In this zone, there are normally no solid particles from the batch and no condensation yet. This "hot zone" is then mainly characterized by high temperatures and soda vapors. The choice of refractories is less critical and several materials can perform properly. For the walls, the main point to focus on is hot crushing strength and creeping resistance, resistance to corrosion is less important as the walls are protected by the checkers from direct contact with the hot fumes. High quality sillimanite bricks are a good choice but silica is also suitable in many cases. The checkers should resist to alkaline vapors (soda) this is why silica-containing refractories (AZS, Sillimanite) must be avoided as they can critically react and form, for instance nepheline. At some extent, basic refractories with C_2S as a binding phase can also be affected. A generation of low binding phase content (direct-bonded) is a possible solution. Finally, in this area also, fused cast alumina based checkers have a solid reputation of durability.

The condensation zone (800-1150°C) is the area where the vapors' species will condense into solid state. The relatively cold temperature makes the choice for wall materials not particularly critical. Generally speaking, Sillimanite or Silica are the typical choices for this part, the point is to support the weight of the structure above without distortion. The situation is absolutely different with regard to the checkers' choice. This part is very sensitive to corrosion and plugging by condensates.

We mainly observed two situations. The first and, by far, the most common, is the formation of solid Sodium Sulfates from a combination of Soda and Sulfur vapors. The result is the formation of large concretions that may plug and corrode the checkers. The use of Fused Cast checkers (AZS or Spinel) with no open porosity helps to both increase corrosion resistance and also limit the sticking of these condensates. Fused Cast checkers also better withstand the thermal cleaning when this emergency maintenance is operated. The magnesite bricks could react with Sulfates to form $MgSO_4$. Another situation is pretty rare but might be unfortunately critical. In case of primary measure for NOx reduction (combustion air decrease with possible reducing conditions or even the 3R process), the corrosion mechanism will be changed into something much more aggressive. The possible condensation of pure caustic soda (NaOH) has been reported. Sodium Hydroxide corrosion is fast and critical for most of the common refractories on the market. A few furnaces, operating very strong reducing conditions (with CO% in the fumes higher than 5 000ppm) have observed a total collapse of checkers within 3 or 4 years. The safest recommendation today is to avoid long production time in such combustion settings.

A new Fused Cast material (pure fused Spinel) is available on the market since 2010 with very promising laboratory and industrial results in such conditions. The very bottom part of the regenerator (below 800°C and basement) is normally not exposed to any corrosion mechanism. The situation here is a combination of thermal cycling and mechanical load. The last courses of checkers must offer a proven resistance to the thermal shock as the bottom part is where the amplitude between cold air and hot fumes is maximum at the inversion (about 500°C). This is also true for the rider arches and the transition bricks where Sillimanite (60% Alumina) represents the large majority of the furnace worldwide. Some lower qualities like Fireclays (40% Alumina) are still used here and there but may not resist in case of thermal cleaning with burners at the bottom. Figure 9 below describes the SEFPRO recommendation for checkers materials.

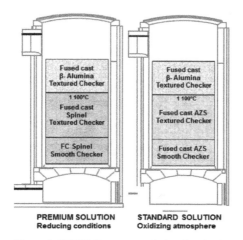

PREMIUM SOLUTION   STANDARD SOLUTION
Reducing conditions   Oxidizing atmosphere

Figure 9. SEFPRO recommendation for checkers

CONCLUSION

When repairing or designing a new soda lime furnace, the regenerator part should not be underestimated. More and more, issues of regenerators are the limiting factors for their lifetime. Even if the regenerator does not directly impact the glass quality, its stable and reliable operating condition is essential for a good control of the melting. Moreover, the efficiency of the regenerator will directly impact the operating cost and, in other words, the competitiveness of the producer.

The regenerator must be carefully dimensioned to offer the best investment/performance ratio. To do so, an adequate volume of chamber is required with a suitable repartition between section and height. Some guidelines have been discussed in this paper targeting particularly fumes and air velocities.

The checkers themselves are of course very important also. The use of textured checkers with open flues pattern (cruciforms Type 4 for instance) offers a significant differentiation with the other conventional checkers.

Finally, to maximize its lifetime, an adequate choice of refractory materials has to be done. For the chamber walls and crown, the use of mullite at the very top and high quality Sillimanite for the middle and bottom areas represents a safe approach.

Sensors and Control

DETECTION OF EARLY STAGE GLASS PENETRATION AND WEAK REFRACTORY SPOTS ON FURNACE WALLS

Yakup Bayram[1], Alexander C. Ruege[1], Eric K. Walton[1], Peter Hagan[1], Elmer Sperry[2], Dan Cetnar[2], Robert Burkholder[3], Gokhan Mumcu[4], Steve Weiser[5]

[1] PaneraTech, Inc., Chantilly, VA
[2] Libbey Glass, Toledo, OH
[3] The Ohio State University, Columbus, OH
[4] University of South Florida, Tampa, FL
[5] Owens-Illinois, Perrysburg, OH

ABSTRACT

Erosion of the refractory lining in molten glass furnaces is a major problem for the glass manufacturing industry. When erosion on the walls is not detected early enough, it may lead to a molten glass leak through the refractory lining and may result in the suspension of production for several weeks. In some cases, a catastrophic accident may also result. The glass penetration typically starts small within the insulation layer and takes anywhere from a few weeks to several months to penetrate through the insulation layer and result in major catastrophic furnace leak. Therefore, detecting an early stage glass penetration within the insulation layer and identifying weak refractory linings will result in safer and longer furnace operation through preventive and proactive maintenance.

To address this major industry need, we are developing a non-destructive sensor technology for tomographic imaging of insulation and refractory lining. This sensor will identify early stage glass penetration into insulation and identify weak refractory spots for preventive and proactive maintenance. We have already developed a sensor that measures the residual AZS thickness on operational glass furnaces. We have also shown the feasibility of mapping interior walls of insulation layers for glass penetration in an operational furnace. Lastly, the same sensor technology is capable of detecting voids and defects in cold refractories.

In this paper, we will discuss the underlying fundamentals behind the proposed sensor technology, the measurement results pertaining to feasibility and in-situ tests on operational furnaces, and the path forward to an integrated sensor system for smart (self-sensing) furnaces.

INTRODUCTION

Erosion of refractory lining of glass furnace walls is a major problem for the glass manufacturing industry. When erosion on the walls is not detected early enough, it may lead to a molten glass leak through the refractory lining and may result in the suspension of production for several weeks. Detecting early-stage glass penetration before a major leak occurs will result in safer and longer operation. These glass penetrations can be detected using a wireless non-destructive sensor technology, currently under development at PaneraTech, Inc. The technology is based on radar imaging of the internal wall structures and utilizes specialized imaging and mapping techniques. The hardware is designed to have a very low profile, allowing it to fit close

157

to the furnace walls and out of the way of any structural or cooli ng elements. In this paper, w e will discuss the underlying fundamentals behind the proposed sensor technology, the measurement results pertaining to feasibility, and the path forward to an integrated sensor system for smart (self-sensing) furnaces.

TOMOGRAPHIC SENSOR FOR EARLY STAGE GLASS PENETRATION DETECTION

The furnace tomographic sensor maps and identifies early stage glass penetration into the insulation layers backing the refractory lin ing of glass furnaces. This s ystem is based on radar imaging of the insulation layers. The sensor is comprised of an ultra-wideband low-profile antenna specially designed to perform tomographic mapping of the insula tion walls. The antenna is connected to radio-frequency (RF) hardware that generate and receive the RF signals into and from the wall. This whole package is designed to fit between structural, cooling, and other elem ents around the furnace and the furnace wall itself and does not touch the wall. The sensor is then scanned over the wall in a two-dimensional pattern. After the data is collected, it is processed and tomographic images of the internal structure of th e wall are constructed. In this way, any glass penetration into the insulation material is identified and mapped.

A conceptual diagram of the scanning system is shown in Figure 1 (a). A common sidewall configuration is shown, consisting of the fused-cast AZS lining, bonded AZS, super-duty firebrick, and fiber board layers. This system will also work with other configurations, such as bonded AZS backing fused-cast behind the fiber board. Only the signal processing changes for the different configurations. The scanning system is configured for the furnace wall so that it will fit around the structural and other components. The system used for the research and development of the sensor itself is shown in Figure 1(b). The system was built to allow for testing on the different sides of the test furnace and at other locations.

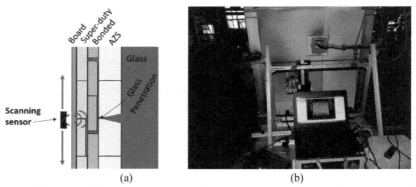

(a)   (b)

Figure 1. (a) Concept of the furnace tom ographic imaging system. (b)Scanning system used for development of the sensor.

To have a fundamental understanding of the materials at high temperatures, we measured the RF propagation loss at high temperatures. The RF properties of the m aterials change as temperature increases. The probe must be desi gned for these properties. These properties were measured in-house at PaneraTech, Inc. using a small kiln to heat the different bricks and measure the RF properties at high temperatures. Using this data, the probe antenna was m odeled using a computational electromagnetic code and designed to achieve the desired performance. Prototype sensors where subsequently built and the feasibility of these was shown and is detailed in the next section.

GLASS PENETRATION MEASUREMENT STUDIES

To demonstrate the feasibility of the furnace tomography sen sor, we tested the system on a developmental furnace at Libbey, Inc. in Toledo, OH. The furnace had four sidewalls consisting of AZS. On one sidewall, a bonded AZS, super-duty firebrick block and fiber board were placed against the AZS. The bonded AZS and super-duty block had pre-cut channe ls to allow glass to flow from a hole drilled in the fused-cast AZS to the outer layers. These bricks where surrounded by the same type of brick, but without channels. The test furnace at Libbey is shown in Figure 2. A drawing of the chann els cut into the layers ar e also shown. A vertical and horizontal channel guide the molten glass into the horizontal groove cut into the super-duty brick. Mortar was used to seal these blocks to make sure the glass was contained in this small area. The installation of the bricks against the fused-cast AZS are also shown in Figure 3.

We performed the tomographic mapping over the area with these cut-ou ts in addition to the surrounding wall that does not contain glass pene tration. This area is shown in Figure 4. A thermal image of the outside of the wall is also shown. There is no indication of glass penetration in this image: the outer surface of the panel is relatively u niform in temperature. The internal furnace temperature was held at 2500 °F during these tests.

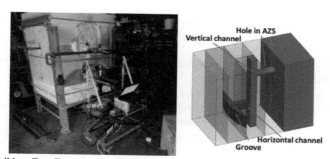

Figure 2. Libbey Test Furnace and diagram of the channels created inside the insulation wall to create glass penetration.

Figure 3. Photographs of the cut-outs into the super duty and bonded AZS bricks and drilling the fused-cast AZS.

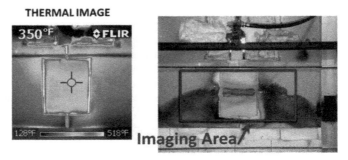

Figure 4. Thermal image of the area with glass penetration and the furnace tomographic system imaging area on the test furnace.

After signal processing, the tomographic images are formed of the internal wall structure. Due to the different material properties of the bricks, we expect a signal to be measured that correspond to the interface between each brick. Any glass penetration into this layer will alter this signal, and depending on the material properties of the blocks, the signal may increase or decrease in power. These effects are readily identified in the imaging results in Figure 5. We show horizontal vs. depth slices in the tomographic imagining results. The first slice is a cut throug h the vertical center of the blocks. We observe signals at each interface along the horizontal scan, except in the center of th e scan at the fused-cast/bonded AZS in terface near 9 inch depth. This lack of signal indicates that glass penetration has occurred at this lay er. The second slice is cut through the groove in the super-duty block. The shape and location of groove is very well defined in the imaging in this slice. This indicates that we can easily map glass that has eroded into joints and into the blocks, not only at just the brick interfaces.

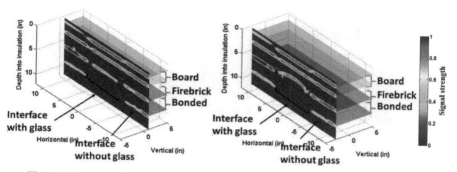

Figure 5. Tomographic imaging results at two different vertical levels inside the blocks.

Given these known signatures from the glass penetration at each depth, it is then possible to automatically identify any glass penetration into the la yers. W e developed a n automatic detection algorithm and applied this to the collected data on the test furnace. The results are shown in Figure 6 below. This glass identification is shown at the fused-cast/bonded AZS interface. As we see, glass is only identified in the 10" wide panel. No glass is found in the surrounding areas at that interface. It is also found inside the gr oove in the super-duty block as well. Once these areas have been identified, they can be marked, reported and visually displayed.

The test furnace was shut-down and after cooling, the sidewall was opened to con firm glass penetration into the channels and interfaces. Figure 7 are photographs showing that the glass penetrated to the super-duty groove. It was also found that glass had filled the interface between the bonded AZS and fused-cast AZS. As noted above, the presence of the glass at this interface was determined due to the change in the signal received at this interface. These results prove the feasibility of the sensor technology for glass penetration detection.

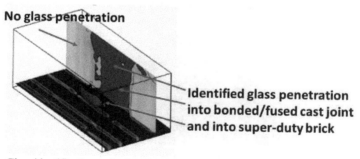

Figure 6. Glass identification at the fused cast/bonded AZS interface joint and into the groove in the super-duty firebrick.

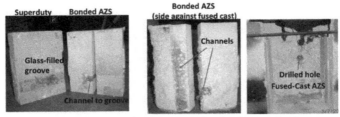

Figure 7. Glass penetration into channels after furnace shutdown.

INTEGRATED SELF-MONITORING SMART FURNACE TECHNOLOGY

PaneraTech has already developed technology for probing exposed AZS refractory lining thickness. This has been im plemented in a hand-held solu tion and an installed, in-situ solution. These sensors work in contact with the exposed AZS wall. The hand-held version allows the user to walk around the furnace and probe as many spots as possible. The in-situ version is installed on the furnace itself to continuously and automatically monitor refractory thickness. Two in-situ probes have already been installed at Libbey 's Toledo, OH plant on an operational furnace. The control box installed near the furnace is shown in Figure 8. The probes are only in contact with the wall during a measurement (which can be less than a second) and automatically retract when not performing measurements. Any number of in-situ probes can be installed in hard-to-reach areas such as throat and in areas where constant monitoring is required. These can be used to monitor critical locations such a s the furnace throat and electrode blocks. Currently, these probes can measure the thickness of the AZS lining up to 5 inches. Work is underway to further increase the measurable range of these probes.

Figure 8. Control box for in-situ probes installed at Libbey, Toledo, OH.

These in-situ AZS thickness sensors, with the hand-held sensor, can be integrated together with the furnace tomography sensor system to create a self-monitoring smart furnace technology to avoid costly leaks and optimize furnace campaign. All furnace health and statu s data can b e combined to generate reports and send alerts. The operators and maintenance staff will thus have awareness of the health of the furnace and be able to observe trends over time as well. This awareness will result in safer and longer furnace operation of furnaces through preventive and proactive maintenance.

CONCLUSIONS AND FUTURE WORK

We have proven the feasibility of the furnace tomography system on the Libbey test furnace. Using the small furnace at 2500 °F internal temperature, we were able to map and automatically identify glass penetration at the bonded AZS/fused-cast AZS interface and into the super-duty block as well. Further refinement of the sensor hardware and image processing is underway using a new test furnace at Libbey in Toledo, OH. This new furnace has several different walls that emulate several different kinds of glass penetration scenarios observed in actual glass furnaces. Additionally, this test furnace will enable further development of the system to measure and image the glass interface behind the insulation layers and AZS lining. Further steps include installation of the low-profile scanning system and demonstration on an operational furnace at Libbey, Inc. We will also further develop the signal processing to allow for imaging through extruded metal gratings that are used on some furnaces in addition to mapping behind structural beams located against the sidewalls.

Given the compact and flexible nature of this system, the full sidewall area of a furnace can be monitored allowing for glass penetration and refractory lining weak spots to be detected and monitored over time. Integrating this system with PaneraTech's other furnace monitoring solutions, a self-monitoring smart furnace can be realized.

FAST AND OBJECTIVE MEASUREMENT OF RESIDUAL STRESSES IN GLASS

Henning Katte
ilis gmbh
Erlangen, Germany

ABSTRACT

Mechanical stresses can strongly impair the fracture strength and processing ability of glass products such as bottles, food jars, tableware, and float glass. Testing for residual stresses close to production is therefore an important constituent of quality control. For decades manually operated polariscopes and polarimeters have been the standard method for testing the level of residual stress in glass. Polariscopes visualize stress by creating false colors that can be visually compared to reference standards (strain discs or retardation scales) in order to determine the magnitude of stress, but this method is rather qualitative than quantitative and does not work for colored glass. Polarimeters on the other hand allow a quantitative measurement by determining the stress-induced polarization change with a rotatable analyzer. However, the measuring results obtained with manually operated polarimeters are strongly dependent on the operator and therefore subjective. In addition, colored glass is hard to measure since the intensity of the light source is often not sufficient for a reliable measurement. The results of statistical Gage R&R tests show that the reproducibility achieved with this method is not acceptable. A family of newly developed imaging polarimeters features the objective measurement of inherent stresses in glass – as random sample test or directly in the production process. The instruments are capable of measuring and visualizing stresses in glass in real time and can be used flexibly wherever conventional polariscopes and polarimeters are still used nowadays. However, in contrast to these, they deliver objective and reproducible values; the influence of different operators on the measuring result is thus largely eliminated. A comprehensive Gage R&R study shows substantial improvements in respect to absolute accuracy and practical reproducibility of the measurement.

INTRODUCTION

Glass is optically isotropic in the relaxed condition, i.e. the refractive index is equally large in all spatial directions. However, mechanical stresses induced by material or production lead to deformations in the material structure and thus to different particle densities along axes. As the propagation velocity of light depends, among other things, upon the density of the material, this sort of change in the microstructure leads to different velocities of light in the medium and thus to a direction-dependent change of the refractive index. The medium therefore becomes birefringent under stress; one designates this effect as stress birefringence (SBR).

In principle, one can measure stresses by determining the propagation velocity of light along different axes. The differences occurring in this case are a direct measure of the birefringence and thus of the stresses causing it. However, instead of determining the propagation velocities or the resulting phase difference directly, e.g. by interferometry, one can use the photoelastic effect to directly induce changes in polarized light.

Linearly polarized light, its electrical field oscillates only in one plane, can be decomposed graphically as a superposition of two collinear light waves oriented at an angle of 45° to the original light wave, of fixed relative phase and at right angles to one another. If these two waves propagate at the same speed, the peaks and valleys of both waves coincide with one another. The addition of the waves produces the original, linearly polarized light wave (Figure 1(a)).

However, if the two light waves are propagated at different velocities within a birefringent material of given length, a delay is induced between the two waves which is designated as optical retardation and is stated in nm or in fractions of the wavelength of light. If the retardation is exactly one quarter of the wavelength, the resulting field vector describes a circle, and one speaks of circularly polarized light (Figure 1(b)). However, the retardation is generally not equal to a quarter of the wavelength and elliptically polarized light is the result. Linear and circular polarization can also be understood as special cases of elliptical polarization.

Linearly polarized light entering a birefringent material therefore leaves this graphically as a superposition of two light waves oriented with polarizations at right angles to one another but now with differing phase, so that in the general case elliptically polarized light will be present. The ellipticity of the emerging light, i.e. the ratio between the ellipse axes, is in this case a measure of the birefringence and thus also of the inherent stresses in the material.

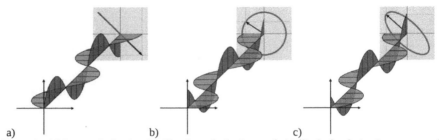

a)   b)   c)

Figure 1. a) Linear polarization, b) Circular polarization, and c) Elliptical polarization, expressed as superpositions of two, perpendicular waves with different retardations

PRINCIPLE OF MEASUREMENT [1]

Instead of measuring the shape of the ellipse directly, one can convert the elliptically polarized light back to linearly polarized light with the aid of a quarter-wave plate (QWP). The method is designated as the compensation method according to de Sénarmont; the associated measuring setup is shown in Figure 2. The quarter-wave plate transforms the elliptically polarized light back to linear polarization, but the polarization direction thereby is changed by a certain angle that is proportional to the retardation and thus also represents a measure of the ellipticity as well as of the birefringence. The angle can be determined in a simple manner by rotating a second polarizer (named analyzer) until an intensity minimum is reached for the viewed measuring point.

In the initial position the analyzer is arranged at right angles to the polarizer. Without a birefringent sample, one therefore obtains a dark image as the light is absorbed completely by the analyzer. However, if one introduces a transparent test specimen with, for instance, edge stresses running tangentially, then these lead to local bright regions (Figure 3 on the left), as the material stresses in the diagonals are at 45° to the orientation of the polarizer axes and thus a part of the light is let through by the analyzer. If one now rotates the analyzer, the intensity changes (Figure 3 center and on the right). The minimum intensity is reached when the polarization direction of the analyzer is again at right angles to the polarization plane of the light.

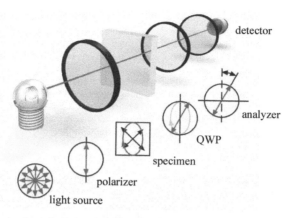

Figure 2. Schematic of a polarimeter for measuring stress birefringence according to the Sénarmont method

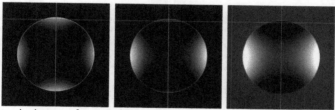

Figure 3. Intensity images of a strain disk with pre-defined edge stresses under different analyzer positions. The polarization plane lies diagonally to the figure axes

The optical retardation in nm as measure of the (stress) birefringence can be determined by multiplying the determined angle by the illumination wavelength and dividing by 180 degree. If the birefringence is homogeneous along the optical path, the retardation can be normalized by dividing through the sample thickness. For describing the magnitude of residual stress in container glass, ASTM C 148[2] also defines the unit of apparent temper number (ATN). The definition is based on the nominal retardation of a Strain Disc standard that was originally used for visual comparison with a polariscope. One temper number corresponds to a retardation of 22.8 nm. Taking the sample thickness into consideration, the apparent temper number can be normalized to the so-called real temper number (RTN). By definition, both values are equal for a sample thickness of 0.16 inch (4.06 mm).

CONVENTIONAL MEASUREMENT

Up to now, manually operated polariscopes and polarimeters have been the standard method for testing the level of residual stress in glass, e.g. according to ASTM C 148. Polariscopes visualize stress by creating false colors that can be visually compared to reference standards (strain discs or retardation scales) in order to determine the magnitude of stress, but this method is rather qualitative than quantitative and is not suitable for dark and colored glass.

Polarimeters on the other hand allow a quantitative measurement by determining the stress-induced polarization change with a rotatable analyzer as described before. However, the measuring results obtained with manually operated polarimeters are strongly dependent on the operator and therefore subjective. In addition, colored glass is hard to measure since the intensity of the light source is often not sufficient for a reliable measurement.

Figure 4 shows a pharmaceutical vial with circumferential stress rings as observed in a conventional polariscope and a conventional polarimeter, respectively. In the polariscope image, areas with high residual stresses appear in yellow and blue color in the normally magenta-colored field of view. The different colors correspond to tension and compression and the color intensity can be used to estimate the magnitude of stress. In the polarimeter image, high residual stresses appear as bright areas in the normally dark field of view. Note that reflections at the vial sidewalls also change the polarization state and lead to false signals.

Figure 4. Pharmaceutical vial in a conventional polariscope (on the left) and in a conventional polarimeter (on the right)

The disadvantages of the above, still largely visual method include the difficulty of determining the position of the intensity minimum with the required accuracy and reproducibility. Additionally, a statement about the spatial distribution is possible only with a large number of measurements and much time. Finally, only stresses that are oriented at 45° to the polarization direction can be measured with this method. All other orientations appear attenuated, as can be seen on the left in Figure 3. The test specimen must therefore be aligned correspondingly before the measurement. As a result, the measuring results obtained with manually operated polarimeters are strongly dependent on the operator and therefore subjective. The results of statistical Gage R&R tests show that the reproducibility achieved with this method is not acceptable.

AUTOMATIC MEASUREMENT

In view of the described difficulties, it makes sense to automate the measuring setup and procedure. An imaging measuring system that facilitates rapid, spatially resolved determination of the magnitude and orientation of stress birefringence with high accuracy has been developed. The applied functional principle essentially corresponds to the setup indicated in Figure 2, but the analyzer is operated by a motor and the human observer is replaced by a matrix camera.

As already mentioned, only stresses that are oriented at 45° to the polarizer axis can be determined using the Sénarmont method. For this reason several measurements are performed under different polarizer settings. The partial results are then combined to an overall result as shown in Figure 5.

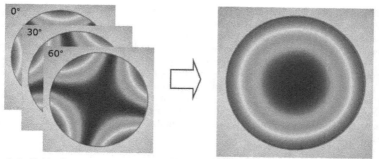

Figure 5. Individual measurement results of a strain disk at different polarizer orientations and the overall result showing the expected radially symmetric stress distribution

MEASUREMENT OF OPTICAL GLASS

Even small local variations of the refractive index can negatively influence the imaging properties of optical components and thus their function. In addition, birefringence alters the polarization state of transmitted light, this being detrimental for applications such as those in metrology. Exact determination of stress birefringence and its spatial distribution is therefore of great importance in the manufacture of optical materials and components.

Figure 6. Automatic polarimeter system for high-resolution measurement of the residual stress distribution in lens blanks with a diameter of up to 300 mm

Solutions have been available for some time now with which the stress birefringence for individual measuring points can be determined automatically with the required accuracy. However, these measuring systems deliver information only for a small measuring area. The test specimen must be scanned in order for larger surfaces to be measured. The spatial resolution achievable in an acceptable measuring time is correspondingly low for large sample dimensions. Furthermore, high requirements are placed on the optical flatness and surface quality of the sample in order to avoid issues with beam deflection and scatter. Through the use of a matrix camera instead of a photo detector, the developed imaging polarimeter can analyze and entire measuring field in a single shot, obviating the need to scan the sample and repeat single point measurements. The lateral resolution is determined in this case by the size of the measuring field and the resolution of the camera.

OPTIMIZATION OF THE ANNEALING PROCESS

When container glass is formed, strong mechanical forces are generated in the glass due to the fast cooling rate. These must be reduced in the annealing lehr by defined heating up and slow cooling down. The energy consumption necessary for this can be significantly reduced by selective optimization of the lehr settings. The annealing lehr is usually operated in a safe range, so that the glass is relaxed more than would actually be necessary. Accordingly, more energy than necessary is frequently expended.

In the case of container glass, usually only the base is measured, since experience has shown that the residual stresses are most critical there due to the contact with the conveyor belt. To enable non-destructive measurement, the detector head is automatically moved close to the neck finish for this purpose (Figure 7). Figure 8 shows a typical measuring result for the base of a food jar.

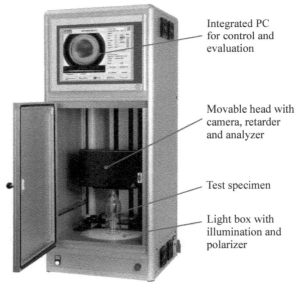

Integrated PC
for control and
evaluation

Movable head with
camera, retarder
and analyzer

Test specimen

Light box with
illumination and
polarizer

Figure 7. Automatic polarimeter system for measuring annealing stress in bottles and jars

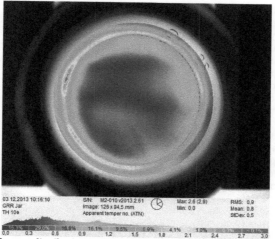

Figure 8. Residual stress distribution in the base of a container glass. For historical reasons, the measured values are usually displayed in the unit of apparent or real temper number

The automated measurement achieves a high practical reproducibility of the measurement in the range of 1/10 temper number, that allows an accurate differentiation of the results depending on the position in the annealing lehr and in the IS machine (Figure 9).

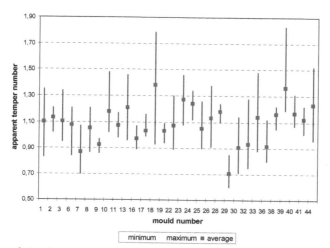

Figure 9. Correlation between the residual stress level and the position in the IS machine. The boxes show the mean values by mold number, the lines represent the min-max spread

CORD STRESS MEASUREMENT

The fracture strength of container glasses is strongly influenced by so-called cord stresses. Especially high stress values near the glass surface can significantly increase the probability of breakage when the bottle is filled in the bottling factory or handled by the end-customer. Continuous and objective control of cord stresses is therefore an essential precondition for ensuring high quality.

Since the cord streaks are often hardly visible in a normal polariscope used for checking the annealing stress, especially in dark-colored bottles, a different method is used to quantitatively measure cord stress. According to ASTM C 978 [3], ring sections are cut from the cylindrical part of the container and analyzed with a polarizing microscope. In order to avoid time-consuming polishing of the glass surfaces, the ring is immersed in a liquid that has a similar refractive index as the glass, eliminating light scattering. The polarized light sent through the ring section changes its polarization state when interacting with stresses in the glass. The resulting optical retardation can be quantified by using a Senarmont or Berek compensator attached to the microscope.

However, the described method is labor-intensive and time-consuming since the ring section has to be scanned visually through the microscope to search for the maximum stress value. In addition, the results very much depend on the operator's skills and are therefore subjective. Gage R&R studies have shown that the practically achievable reproducibility is significantly below the desired level.

In order to make the measurement of cord stress more objective and reproducible, the automatic polarimeter has been further developed and adapted. It applies the same physical principles as the manual method, but fully automates the measurement of cord stresses in bottles and jars with a diameter of up to 120 mm. After placing the ring section into a petri dish with immersion liquid, the operator only has to start the measurement. After less than one minute the operator sees a color-coded image showing the stress distribution for the whole ring section. In addition, the area of maximum tension is identified, highlighted in the image and reported as a numeric value (Figure 10).

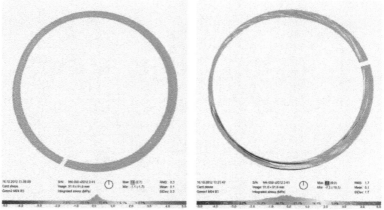

Figure 10. Measurement results of a bottle ring without stress (on the left) and a sample with significant cord stress (on the right). Compression is shown in blue and tension in red. Neutral areas appear in green

MEASUREMENT OF TUBULAR GLASS

Since the measurement is limited to two dimensions, the optical retardation along the optical path is integrated. I.e. if compression and tension is present at the same time in different layers of the glass, only the difference can be measured. For cylindrical objects such as tubes, vials or syringes, it is possible to measure the stress distribution in an axial cross-section non-destructively by immersing the specimen into an index-matching liquid and measuring tangentially through the sidewall (Figure 11).

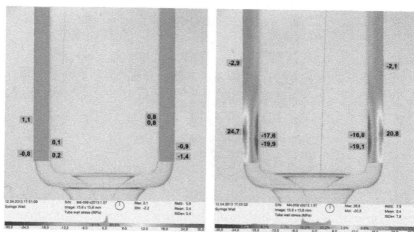

Figure 11. Measurement results of an annealed and a non-annealed syringe. Compression is shown in blue and tension in red. Neutral areas appear in green. The peak values at the glass surface are labeled accordingly.

MEASUREMENT IN REAL-TIME

The previously described measuring procedure requires rotating optical elements which requires time and is therefore not suitable for moving objects. The measuring time of typically 30 seconds also limits throughput, so that the test is normally made in the form of random sampling downstream from production. To accelerate the measurement and enable a higher testing frequency, the mechanically operated analyzer is replaced by a polarization camera that delivers multiple images with different polarizations at the same time. A computer calculates the stress distribution from these images in real-time. Figure 12 shows the developed system.

The high frame rate of up to 20 Hz makes it possible to monitor dynamic processes and enables production to be 100% tested (Figure 13).

Figure 12. Automatic polarimeter system for measuring the stress distribution in real-time

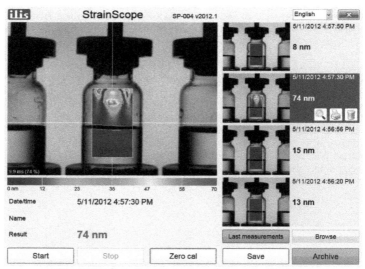

Figure 13. Inline stress inspection of pharmaceutical vials

GAGE R&R ANALYSIS

In order to determine the reliability of the three measuring methods (manual with a conventional polariscope as used in Figure 4, fully automatic with the automatic polarimeter system as shown in Figure 7, and semi-automatic with a real-time polarimeter system as shown in Figure 12) a Gage R&R (Gage Repeatability and Reproducibility) study has been conducted. For this purpose, a set of eight food jars and a set of twelve beverage bottles with different stress levels were measured with each setup. Each sample was measured three times by three different operators, resulting in a total number of 72 and 108 measurements, respectively per instrument.

A measure for the suitability of a testing method is the Total Gage R&R percentage value which is calculated from the measurement results. A high value expresses a large operator influence and thus a low reproducibility. By definition, Gauge R&R values below 30% are considered as acceptable, values below 20% are good and values below 10% are excellent. Table 1 shows the overall results. Figure 14 and 15 show the interval plots for each sample set.

Table 1. Total gage R&R values achieved with different gages

Gage	Bottles	Jars
Conventional polarimeter	31.8%	60.9%
Automatic system (StrainMatic)	4.3%	3.4%
Real-time system (StrainScope)	10.9%	11.7%

While all gages delivered comparable average values, the variance of the conventional polarimeter is much higher and the resulting Gage R&R value exceeds the threshold of 30% significantly. The fully automatic system performs best, since the sample handling with the real-time system (i.e. adjusting the region of interest within the camera field of view) is still manual and thus influenced by the operator.

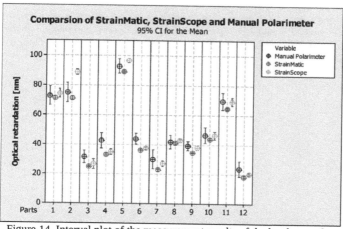

Figure 14. Interval plot of the measurement results of the bottle samples

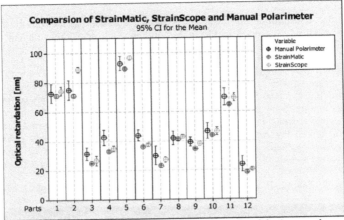

Figure 15. Interval plot of the measurement results of the jar samples

CONCLUSION

The automatic measurement offers numerous advantages in comparison to the manual measurement with conventional polariscopes and polarimeters. Besides an increased level of reliability through a much better repeatability and reproducibility, the developed systems allow it to file the measuring results for documentation purposes and subsequent statistical analysis. The real-time capability makes it possible to integrate the technology into the production processes, e.g. into inspection machines.

REFERENCES
1.   H. Katte, Imaging Measurement of Stress Birefringence in Optical Materials and Components, Photonik International 2009/1, p. 39
2. ASTM C 148 – 00, Standard Test Methods for Polariscopic Examination of Glass Containers
3. ASTM C 978 – 04, Standard Test Method for Photoelastic Determination of Residual Stress in a Transparent Glass Matrix Using a Polarizing Microscope and Optical Retardation Compensation Procedures

FEEDER EXPERT CONTROL SYSTEM FOR IMPROVED CONTAINERS

Fred Aker
Nikolaus Sorg GmbH & Co. KG
97805 Lohr am Main
Germany

ABSTRACT
In recent years numerous efforts have been undertaken in container glass production to achieve better control and to improve production efficiency. This is especially true in regard to light-weight glass containers. This in particular requires stable and reproducible conditions in the glass conditioning systems. Repeatability is crucial, especially the ability to return to known good operating conditions quickly following frequent job changes. SORG and Glass Service are employing Expert Systems to automate and capture best practices on glass container forehearths. This paper present results and ongoing investigations to improve and automate these critical forehearth processes.

INTRODUCTION

Advanced Expert Systems such as *ESIII*™ have been used successfully for furnace control for two decades. Advantages include predictability as well as the chance to bias the energy balance to the cheaper media. In most geographical areas this means more fossil energy and less electrical boosting. This results in lower energy costs, and often less total energy consumption.

ESIII™ has also been used successfully in continuous operation channels such as fiber glass. Up until now, there has been a reluctance to employ Expert Systems on highly variable container glass forehearths. Highly variability is due to the tendency for shorter runs and the need for flexibility. Flexibility meaning large pull changes on the furnace and individual forehearths. This paper explores the possible benefits of employing an Expert System on container glass forehearths against the perceived risks.

Expert systems are widely accepted in modern furnace operations. Figure 1 summarizes the benefits of the expert system. The benefits have been presented at various symposiums including the Glass Problems Conference. Advantages include:

- Self-learning systems that mimic the operation of the best operator. Insuring reproducible best practices across all shifts.

- Ease of operation: Operators can choose new set points and the Expert System will predict when the new set point will be achieved and show the path to the new set point. This avoids wild temperature swings through under and over correcting.

- More stable operation: Through avoiding the temptation to 'chase' temperature set points, variability in the furnace can be greatly reduced. By up to 80% as reported to the 74th Conference on Glass Problems [1].

- Energy cost savings by using an optimal mix of fossil and electrical energy inputs. The expert system is more willing to operate on the 'bleeding' edge. Operators often add boosting to improve a specific situation and then leave it in place after the underlying problem no longer exists. Why risk reducing boosting when the furnace is producing high quality?

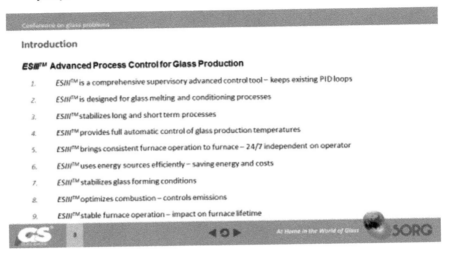

Figure 1. Summary of expert system advantages

The authors believe that the potential impact on glass production is equally high when using Expert Systems on container glass forehearths.

FOREHEARTHS

Modern forehearths are a collection of independently controlled zones. The PID goal is to meet set points in each individual zone. This is done through a combination of fossil fuel firing and possibly electrical boosting. Along with indirect and direct cooling. The ultimate goal is to meet a specific temperature and high thermal homogeneity at the spout. Figure 2 shows the details and complexity and Figure 3, the set point control.

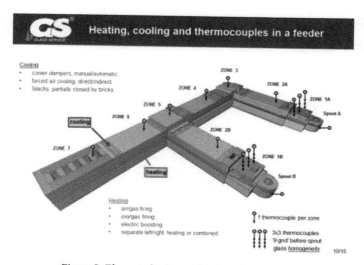

Figure 2. Elements leading to forehearth complexity

The approach of *ESIII*™ is a holistic approach, shown in Figure 3. It is to treat the forehearth as a complete system taking into account production needs as well as what is happening upstream of the individual forehearth.

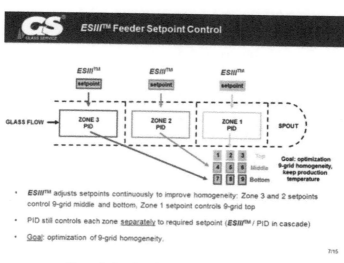

Figure 3. Standard forehearth set point control

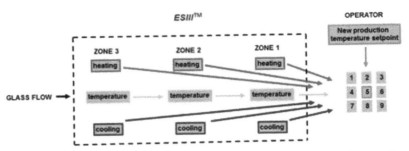

- Operator only needs to specify new production temperature for the article (9-grid or spout).

- *ESIII*™ controls simultaneously all heatings and coolings of entire feeder in one controller, also during the job changes, to reach required new setpoint at optimal homogeneity.

- *ESIII*™ takes into account variations in upstream temperatures and various residence time of glass in feeder at different pulls to control 9-grid temperatures

9/15

Figure 4. *ESIII*™ holistic approach

WHY INCORPORATE FOREHEARTHS?

The major drivers are to increase production and flexibility. More frequent job changes and shorter runs are the new norms. Often SORG supplies forehearths that we feel should be run in automatic mode. When doing audits, we often find these forehearths in manual control mode with heating and cooling setups that we feel are suboptimal. Modern forehearths are extremely flexible. With flexibility, comes complexity. With the qualification of plant personnel sinking globally along with high turnover, these challenges need to be attacked in a new manner. In a manner that reduces the variability among operators of differing skill and experience levels. Figures 5 summarizes the reasons to use the expert system.

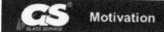

Motivation

- Glass producers answer: NO, we already have good operators

- Mission impossible: feeder adjustment is too complex, too many different situations, take into account influence from furnace temperatures, furnace pull and job changes on neighbor feeders, changes of spout ring, upsets in feeder during job change etc.

- Reality: operators are 'fighting' to setup correct feeder temperatures, 'stress' in the control room with machine operators -> potential need for a helpful tool

- Motivation: lets try it and find a control strategy using *ESIII*™ advanced control to setup feeder fully automatically, including the job changes.

- Goal: operator only needs to set up the final production temperature at the feeder end (9-grid or spout), the rest of the feeder setup is done by *ESIII*™

ESIII™ Control Of Feeder Job Change should setup feeder optimally (glass quality) and efficiently (save energy), in minimal time after job change.

3/15

Figure 5. Summary of reasons to use expert systems on forehearths.
Courtesy of Dr. Menno Eisenga 1st Joint Meeting of DGG – ACerS GOMD – Aachen, Germany 2014

OBSERVED BENEFITS IN LONG-TERM TRIALS

- Easier job changes. The operator can set the desired spout temperature and homogeneity. The operator also has the option of picking an article from an *ESIII*™ system library. The Expert System does the rest.

- Faster path to optimal glass parameters. Through preconditioning, *ESIII*™ starts the transition early to achieve the new situation when the job change is finished.

- Better glass. Through treating the forehearth holistically, we are not cooling in one zone to achieve a set point and heating in another zone to achieve homogeneity.

- Less wasted energy. *ESIII*™ tries to reduce cooling before increasing heating.

CONCLUSIONS

Despite the observed benefits, the goal of this paper has not yet been achieved to our satisfaction. We are heavy on investigations and light on numerical results. We know we are contributing to achieving stable processes quickly. We know we contribute to reproducing known good conditions. We cannot yet conclusively quantify the benefits of Expert Systems on forehearths for container glass furnaces. The level of interest among top factory management is high. Repeat orders are not uncommon. Up until now it has not been possible to do an apples to apples comparison in a controlled before and after experiment where the only variable is the support or lack thereof of an Expert System. Reasons why there is no hard data include:

- The mix of products of products being made is constantly changing.

- The mix of products that were being made prior to the job change varies.

- Incomplete data concerning historical runs.

- Job runs not being run long enough to produce stable data.

- Manual intervention from well-meaning operators.

REFERENCE
1. Erik Muijsenberg, Robert Bodi, Menno Eisenga, Glenn Neff, "An Advanced Expert Control System and Batch Imaging Software for an Improved Automatic Melter Operation, 74th Conference on Glass Problems, 118 (2013).

Modeling

3-D TRANSIENT NON-ISOTHERMAL CFD MODELING FOR GOB FORMATION

Jian Jiao, Oluyinka Bamiro, David Lewis, and Xuelei Zhu
Owens-Illinois, Inc.
One Michael Owens Way, Perrysburg, OH 43551

ABSTRACT

To achieve a quality final product in the glass container industry, it is critical to determine the "ideal" glass gob shape to be produced from the feeder after shearing. Large deviations from an ideal gob shape may result in severe consequences for the gob delivery system and molds. The formation of ideal or desired gob shape is dependent upon operational parameters such as glass temperature/viscosity, uniformity, plunger stoke and heat-loss management. A Computational Fluid Dynamics (CFD) model provides an efficient and cost effective way of studying the effects of these parameters when optimizing gob shapes that are subject to the aforementioned operating parameters and conditions.

For the current study, two CFD approaches were used to create a 3-D transient non-isothermal CFD model in order to study the effects of flow and the thermal condition of molten glass on gob formation.

In the first approach, a numerical model was developed by utilizing the ANSYS POLYFLOW solver in conjunction with both the Mesh Superposition Technique and the Lagrangian adaptive remeshing technique to model plunger motion and gob formation respectively.

In the second approach, a hybrid model using both ANSYS FLUENT and POLYFLOW was developed, in order to achieve higher computational efficiency and a reduction in computational time. The hybrid model consists of two parts: (1) the flow and thermal condition of the molten glass is modeled by FLUENT using the moving deforming mesh technique for plunger motion, and (2) the gob forming process is modeled in POLYFLOW by mapping/transferring the glass flow and temperature information from FLUENT. The hybrid model used in the second approach shows significant improvement in computational performance with reasonable accuracy.

1. INTRODUCTION

There is a continual drive in the glass container industry to improve product quality and reduce manufacturing cost to gain market share over other forms of packaging. Production defects are one of the major sources of manufacturing costs in the glass container industry. Products with defects are sent back to the furnace. They must be reheated at high temperatures which increases energy costs.[1]

In order to improve the quality of finished products and reduce the manufacturing cost, it is important to determine, understand and control the factors that cause temperature inhomogeneity in the glass gob in the early stage. Traditionally, gob productions control has been conducted based on the past experience and operator knowledge[1]. Recent advances in numerical techniques and computing capabilities have made the numerical modeling of the gob forming processes feasible.

The commercial software ANSYS POLYFLOW is most widely used within all three of the major glass industries (container, flat, and fiber)[2]. Early simulation works on glass container forming process with POLYFLOW have been done by Hyre, et al.[3-6]. They developed comprehensive simulations of the overall container forming process (forehearth, feeder, gob delivery and IS machine). However, POLYFLOW has the limitation in its computing scalability, and its parallel computing architecture cannot be used in a multi-node cluster based computing architecture, hence requiring significant computational time for realistic industrial applications[7]. Based on the authors knowledge and experience, computation time for a simplified 3-D transient non-isothermal models (in terms of geometry and physics) are on the order of weeks. Consequently, it is impractical to model industrial applications with complex geometry and physics, using the POLYFLOW solver alone, with an acceptable computational cost.

In this study, a 3-D transient non-isothermal glass gob forming model was developed using two computational approaches. In the first approach, the ANSYS POLYFLOW solver is used in conjunction with both the Mesh Superposition Technique and the Lagrangian adaptive remeshing technique. In the second approach, a hybrid model using both ANSYS FLUENT and POLYFLOW was developed. Particular interests are the thermal non-homogeneity prediction in glass gobs and more importantly is the computational efficiency enhancement by the hybrid model.

2. GEOMETRY SCHEMATIC

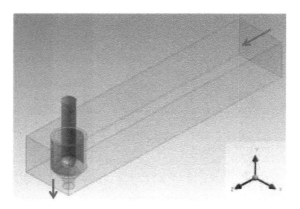

Figure 1: Model Schematic

Figure 1 shows an isotropic view of the feeder system. The feeder system consists of three components: glass tank, plunger and hollow tube. The molten glass enters the glass tank and passes through the feeder channel, taper and then through the outlet of the orifice. A glass gob is formed from there.

The glass tank is 1 m long, 0.2 m wide and 0.15 m high. The total length of the plunger is 0.2 m. The diameter of the hemispherical tip of the plunger is 0.04 m. The inner diameter and outer diameter of the hollow tube are 0.1 and 0.12 m, respectively. The diameter of the taper's large end is 0.08 m, whilst the diameter of small end is 0.04 m. The flute length of taper is 0.03 m. The orifice is 0.03 m long with 0.04 m in diameter.

3. MESH GENERATION

For ANSYS POLYFLOW model, the moving parts, such as the plunger, were taken into account utilizing a Mesh Superposition Technique (MST). The geometries and meshes of the plunger, tube (solid domain) and glass tank (fluid domain) were created separately, as shown in Fig. 2 (a), (b) and (c). The meshes for solid domains (plunger and tube) were then superimposed onto the glass tank domain by using POLYFUSE, a functionality in ANSYS POLYFLOW, as shown in Fig. 3. The total cell count for ANSYS POLYFLOW model is around 31,000.

The advantage of the Mesh Superposition Technique is that the mesh generation is very simple, since no complex intermeshing region between adjacent geometries needs to be generated. In addition, when objects undergo rigid body motion, the Mesh Superposition Technique reduces the computational time since no remeshing is required. However, the Mesh Superposition Technique is diffusive and cannot accurately resolve velocity gradients in the neighborhood of the moving part. In this study, the impact of this disadvantage is limited since the primary interest was the glass gob shape and temperature homogeneity in the gob.

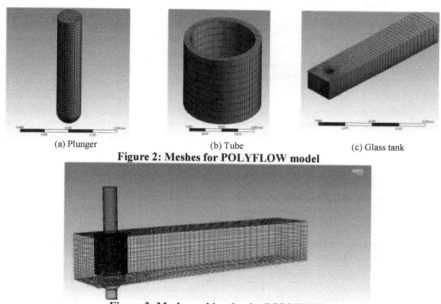

| (a) Plunger | (b) Tube | (c) Glass tank |

Figure 2: Meshes for POLYFLOW model

Figure 3: Mesh combination by POLYFUSE

For the hybrid model (ANSYS FLUENT & POLYFLOW), the moving parts were taken into consideration using a sliding and dynamic mesh model in FLUENT[9]. As shown in Fig. 4, smoothing and dynamic layering methods were adopted to adjust the mesh of a zone with moving and/or deforming boundaries. The total cell count for the hybrid model is 126,000.

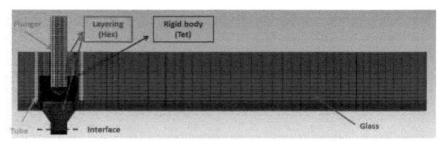

Figure 4: Meshes generated for hybrid model (FLUENT sub-model)

4. MODEL DESCRIPTIONS

4.1 Governing equations
    For the gob forming process, glass was modeled as generalized, Newtonian, incompressible flow. The conservation of mass equation is

$$\nabla \cdot \boldsymbol{v} = 0 \tag{1}$$

where $\boldsymbol{v}$ is the velocity vector.
    The conservation of momentum equation is

$$-\nabla p + \nabla \cdot \boldsymbol{T} + \boldsymbol{f} = \rho \boldsymbol{a} \tag{2}$$

where p is the pressure, $\boldsymbol{f}$ is the volume force. ρ is the density, and $\boldsymbol{a}$ is the acceleration. $\boldsymbol{T}$ is the extra-stress tensor, which is defined as

$$\boldsymbol{T} = 2\eta \boldsymbol{D} \tag{3}$$

where $\boldsymbol{D}$ is the rate of deformation tensor. η is the shear viscosity, which can be a function of local shear rate $\dot{\gamma}$ and temperature T. The local shear rate is defined as

$$\dot{\gamma} = \sqrt{2tr(\boldsymbol{D}^2)} \tag{4}$$

For non-isothermal generalized Newtonian flows, the conversation of energy equation is

$$\rho C_p \frac{DT}{Dt} = r - \nabla \cdot \boldsymbol{q} + \mathbf{T}:\nabla \boldsymbol{v} \tag{5}$$

$$\frac{DT}{Dt} = \frac{dT}{dt} + \boldsymbol{v} \cdot \nabla T \tag{6}$$

$$\boldsymbol{q} = -k\nabla T \tag{7}$$

where $\frac{DT}{Dt}$ is the material derivative of the temperature. q is the heat flux calculated by Fourier's law and k is the thermal conductivity.

4.2 Mesh Superposition Technique

ANSYS POLYFLOW incorporates a method known as Mesh Superposition Technique (MST)[8] to simulate transient flows with internal moving parts. In this study, when the plunger mesh is imposed over the glass mesh as shown in Fig. 3, the conservation of mass, momentum and energy equations (1), (2) and (5) must be modified.

In order to calculate a physically meaningful pressure, even in zones where geometrical penetration occurs, the conservation of mass equation is modified to become

$$\nabla \cdot v + \frac{\beta}{\eta} \Delta p = 0 \qquad (8)$$

where β is a relative compression factor. In this study, the relative compression factor β value is 0.01 by default, which has been shown to be the best choice for this factor in ANSYS POLYFLOW[8].

The conservation of momentum equation is modified to become

$$H(v - \bar{v}) + (1 - H)(-\nabla p + \nabla \cdot T + f - \rho a) = 0 \qquad (9)$$

where H is the step function and v and $\bar{v}$ are the fluid and solid velocities respectively. Eqn. (9) is discretized for each node of the velocity field in the computational domain. The step function H has a value of 0 and 1 for fluid and solid nodes respectively. Consequently, the usual Navier-Stokes equations are recovered from Eqn. (9) for the fluid nodes while Eqn. (9) for the solid nodes degenerates into

$$v = \bar{v} \qquad (10)$$

which indicates the local velocity of moving part $\bar{v}$ is imposed.

The conservation of energy equation is then modified to be

$$(1 - H)\left(\rho_f C_{p_f} \frac{DT}{Dt} - r_f - \nabla \cdot (k_f \nabla T) - T{:}\nabla v\right) + H\left(\rho_s C_{p_s} \frac{DT}{Dt} - r_s - \nabla \cdot (k_s \nabla T)\right) = 0 \qquad (11)$$

where subscript f and s indicate fluid and solid domains respectively. Again, Eqn. (11) is discretized for each node for the temperature field. Similar to the momentum conservation equation, the step function H has a value of 0 and 1 for fluid and solid nodes respectively. When $H = 0$, the energy equation for the fluid domain is solved, while the energy equation for the solid domain is solved for $H = 1$.

4.3 Lagrangian mesh updated and TGrid adaptive remeshing method in ANSYS POLYFLOW

The Lagrangian mesh update method is applied to the 3-D, transient, gob formation process. The Lagrangian mesh update is used for updating the nodes that undergo deformation. In this method, the nodes in the deforming volume simply follow the displacements of the material points in all directions of Cartesian space.

During the gob forming process, the glass volume undergoes large deformation. Consequently, an initial mesh with good mesh quality gradually deteriorates as the glass volume deforms during the simulation process. In order to maintain good quality elements as the simulation progresses and also make mesh quality independent of glass volume deformation history, the TGrid adaptive remeshing method[8] was used for the glass gob mesh. The TGrid adaptive meshing method will mark and re-mesh elements that have become too stretched and/or too distorted due to Lagrangian mesh updates. The elements that have mesh quality metrics lower than user specified threshold ($T_{quality}$) will be marked and refined. The lower the value, the more distorted the element. All of the elements lower than the quality threshold will be re-meshed by tetrahedral (3D) with a user specified remeshing size S.

In this study, remeshing was performed at each time step. The elements of the mesh with $T_{quality} < 0.5$ will be marked for remeshing with tetrahedral meshes having size $S = 1.5$ mm.

4.4 Boundary conditions

For flow boundary conditions, at the entrance of the glass tank, a constant mass flow with uniform velocity in Z direction was specified as

$$\dot{m} = 0.5 \ kg/s \tag{12}$$

The top, side and bottom walls of the tank were specified as a zero wall velocity. The wall boundary conditions for the taper and orifice were also specified as zero wall velocity:

$$v = 0 \tag{13}$$

A free surface boundary condition was imposed on the glass gob surface in POLYFLOW. Since the free surface boundary position is not known in advance, the free-surface problems have additional degrees of freedom and additional equations to solve as compared with fixed-boundary flow problems. For a transient flow problem, the position of moving boundaries is determined by solving a kinematic equation:

$$\left(\frac{\partial x}{\partial t} - v\right) \cdot n = 0 \tag{14}$$

where x is the position vector of a node on the free surface, v is the velocity vector evaluated at free surface and n is the normal vector to the free surface.

Surface tension plays an important role in glass gob free-surface problems. The net influence of surface tension force f_n is in the normal direction and satisfies the following equation

$$f_n n = \frac{\sigma}{R} n \tag{15}$$

where σ is the surface tension coefficient and R is the Gaussian curvature of the surface. In POLYFLOW, surface tension forces are introduced on the right-hand side of the momentum equations.

For the thermal boundary conditions, an insulated thermal boundary condition was imposed on the top wall of the tank. On the side and bottom walls of the tank and the wall of the tapered orifice, the molten glass loses heat energy to surrounding hot air via convection

$$q = h_{c1}(T - T_h) \tag{16}$$

where $h_{c1} = 5$ W/m³/K is the convective heat transfer coefficient and $T_h = 700$ °C is the temperature of the hot air.

Since the glass gob surface is exposed to ambient air with room temperature, the heat flux exchange between the glass gob and air on the free surface was described as

$$q = h_{c2}(T - T_c) \tag{17}$$

where $h_{c2} = 10$ W/m²/K is the convective heat transfer coefficient and $T_h = 30$ °C is the ambient temperature.

4.5 Initial conditions

It is assumed that the tank was filled with molten glass initially at 1160 °C. A semi-ellipsoid volume (the semi-major axis and semi-minor axis are 0.02 and 0.006 m, respectively) of glass was created at the outlet of orifice to simulate the initial gob shape as shown in Fig. 5. The steady state calculation was carried out with the orifice exit modeled as a pressure outlet ($P = P_{atm}$). The results from the steady-state model were then used as initial conditions for the following transient model that simulates the actual glass gob forming process.

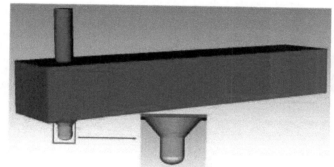

Figure 5: The initial conditions of the model

4.6 Material properties

The viscosity of molten glass displays a broad region of relatively strain-independent viscosity behavior albeit with a high degree of temperature dependence. A number of empirical formula are available to fit the experimentally measured data. In this paper, the Vogel-Fulcher-Tammann (*VFT*) equation[10] is used

$$\log_{10} \eta = A + \frac{B}{T - T_0} \tag{18}$$

where the constants, A, B and T_0, vary according to the actual composition of the glass.

The glass density, specific heat, and thermal conductivity were assumed to be constant since they do not vary much within the gob forming process in this study. The density, specific heat and thermal conductivity of the plunger and tube were also specified as constant values.

Although heat transfer via thermal radiation is vital in determining the temperature profile in the glass flow, it was not taken into consideration in this work, since the main task of this paper is to illustrate the capability and high efficiency of the hybrid model in dealing with the gob forming problems.

4.7 Numerical solutions

Two computational approaches were used to solve the 3-D transient non-isothermal glass gob forming problems. In the first approach, ANSYS POLYFLOW solver was adopted in conjunction with the Mesh Superposition Technique (MST) and Lagrangian mesh update and adaptive remeshing to model plunger motion and glass gob forming.

In the second approach, a hybrid model utilizing the ANSYS FLUENT and POLYFLOW solvers was developed in order to improve the computational efficiency. The entire hybrid model was divided into two sub-models with a common interface (10 mm from the outlet of orifice) as illustrated in Fig. 6. The common interface, henceforward referred to as the interface, will be used for one-way information exchange between FLUENT and POLYFLOW sub-models.

In the FLUENT sub-model, the flow and thermal condition of molten glass is modeled during its passage through the tank to the outlet of orifice. The plunger motion is simulated in FLUENT using the moving deforming mesh technique[9]. The glass flow and thermal conditions ($v(t)$, $T(t)$) at the interface in the FLUENT sub-model will be mapped/transferred to the POLYFLOW sub-model and serve as an inlet boundary condition. The glass flow and heat transfer in a small portion of orifice and gob forming process was simulated in the POLYFLOW sub-model.

For both approaches/models, the total physical time for numerical simulation is 0.3 s. We assume that the glass gob will be sheared off after 0.3 s. Since ANSYS POLYFLOW is not capable of modeling the shearing process, the simulation ends before the glass gob was sheared.

In this study, the time step used in transient calculation is 0.001 s for both ANSYS POLYFLOW model and hybrid model (FLUENT/POLYFLOW). Numerical computations were run with 8 cores on a workstation with Intel Xeon CPUs ES-2640 @2.5GHz.

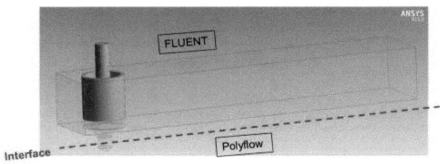

(a) Hybrid model (ANSY FLUENT/POLYFLOW)

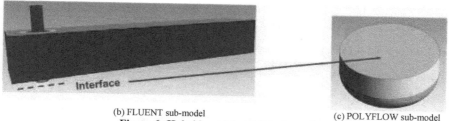

(b) FLUENT sub-model   (c) POLYFLOW sub-model

Figure 6: Hybrid model and its sub-models

5. RESULTS

5.1 ANSYS POLYFLOW model

Numerical simulations were carried out with the POLYFLOW model first. A 3-D, isothermal, steady state case was investigated. In this case, the plunger and hollow tube are static. The molten glass enters the tank channel with uniform velocity in the Z direction and then flows out of the orifice outlet due to the force of gravity. The velocity distribution at steady state in the middle YZ plane (x=0) is shown in Fig. 7 (a). The viscous glass passes through the tank channel with low velocities and then is pulled out from the orifice by gravity at relatively high velocities. Figure 7 (b) shows the 3-D streamlines at the steady state of glass flow in the feeder system. The glass flow becomes fully-developed before it reaches the drain. The tube and plunger block the main flow, which makes it easier to enter the drain from the front than from the rear.

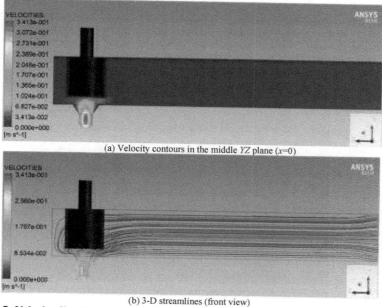

(a) Velocity contours in the middle YZ plane (x=0)

(b) 3-D streamlines (front view)

Figure 7: Velocity distribution for steady state isothermal POLYFLOW model

Figure 8 shows a sequence of 3-D images at various time instances during the glass gob forming process. The plunger moves at a constant speed of 0.04 m/s. The tube is static. The velocity contours are plotted by making the front faces of the post-processed image of the tank transparent in CFD Post. As time advances, the glass gob evolves from a shallow semi-ellipsoid (Fig. 8 (a)) to a cylindrical shape (Fig. 8 (d)). The surface tension and high viscosity of molten glass constrains the movement of the free surface and forms a hemispherical end on the gob.

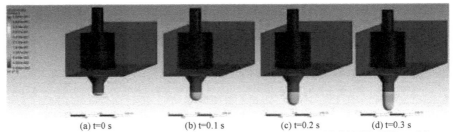

(a) t=0 s   (b) t=0.1 s   (c) t=0.2 s   (d) t=0.3 s

Figure 8: 3-D gob shapes at various instances from isothermal POLYFLOW model

A numerical simulation was then performed for 3-D, non-isothermal, glass gob forming with a POLYFLOW model. Figures 9 (a), (b), and (c) show the steady state temperature distributions on the middle XZ plane, on the XY plane, which is 0.9 m away from the inlet and middle YZ plane, respectively. The temperature of molten glass decreases in the main flow direction due to the heat loss from the side walls of the tank as shown in Fig. 9 (a). The glass temperature close to the top surface is higher than the temperature along the bottom surface as shown in Fig. 9 (b) because the top wall of the glass tank is thermally insulated while the bottom wall experiences heat loss. Symmetric temperature distributions can also be observed in Fig. 9 (a) and Fig. 9 (b) since geometry and boundary conditions are symmetric along the middle YZ plane. However, as shown in Fig. 9 (c), glass temperatures close to the bottom left corner are lower than that near the upper right corner. This non-homogeneous temperature distribution in the feeder system will reduce the thermal homogeneity in the glass gob, as will be discussed in the following section.

(a) Temperature contours on the central XZ plane

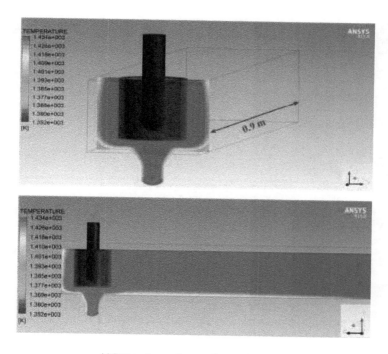

(c) Temperature contours on the central YZ plane

Figure 9: Temperature distribution for steady state non-isothermal POLYFLOW model

Figures 10 (a), (b) and (c) show the temperature contours inside the glass gob on middle XY plane, middle YZ plane and gob shapes with velocities distribution at four different time instances, respectively. As mentioned earlier, the heat losses from the glass to the ambient air are symmetric along the middle YZ plane, but not symmetric along middle XY plane of the gob. Consequently, a symmetric temperature distribution on the middle XY plane inside the gob is observed in Fig. 10 (a). A non-symmetric temperature distribution on the middle YZ plane is also shown in Fig. 10 (b). This non-homogeneous temperature distribution in the glass gob was developed at the beginning (t = 0 s) and then propagated to subsequent times during the glass gob formation, as shown in the Fig. 10 (a) and (b). In addition, since the viscosity of molten glass is sensitive to temperature changes, the non-homogeneous temperature distribution leads to non-homogeneous viscosity distribution, which results in a non-uniform velocity profile in the glass gob. As shown in Fig. 10 (b) and (c), as time advances, the glass gob shape starts to deviate from the cylindrical shape. At the end of 0.3 s, the glass gob tilts at a small angle from the vertical direction. This deviation from the ideal, cylindrically shaped gob will produce adverse consequences in later glass forming procedures.

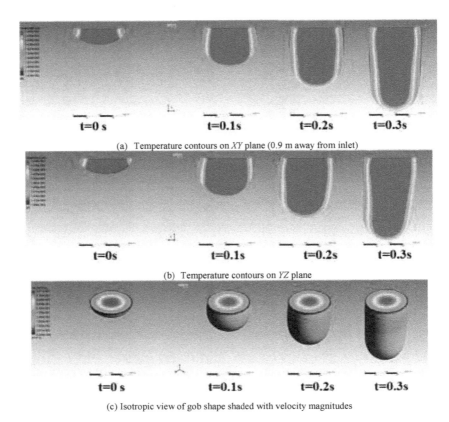

(a) Temperature contours on *XY* plane (0.9 m away from inlet)

(b) Temperature contours on *YZ* plane

(c) Isotropic view of gob shape shaded with velocity magnitudes

Figure 10: 3-D gob formation at four time instances (t = 0, 0.1, 0.2 and 0.3 s) from non-isothermal POLYFLOW model

5.2 Hybrid model (ANSYS FLUENT/ POLYFLOW)

Numerical simulation in glass gob formation provides the advantage and flexibility of investigating effects of both process and geometric factors on glass gob quality metrics when compared to experimental efforts. In addition, numerical simulation provides a wealth of insight to quantitative and qualitative data that are difficult to measure in experiments and/or a glass factory environment. One of the major challenges in numerical simulation of the glass gob formation process is to keep the computational cost (in terms of computing time) to a practical range, given a fixed computing infrastructure within an acceptable level of accuracy. Reducing numerical computation time is crucial in this study in order to reduce the task turn around time and also harness the aforementioned advantages of numerical simulation for glass gob formation. For example, the total computing time in the current study for simulating 0.3 s of the gob formation process using 3-D, transient, isothermal and non-isothermal gob forming model with POLYFLOW are 2.54 days and 16.27 days, respectively. As mentioned in previous section, the

POLYFLOW model has modest cell count of 31,000, much less compared to representative cell count encountered in general industrial applications that possess complex geometry and/or ensuing physics. Seeking an alternative, efficient and accurate way to model 3D gob forming process is necessary.

An alternative method was introduced using a hybrid model. In the hybrid model, a FLUENT sub-model was used to model glass flow and thermal conditions with plunger motion from the entrance of the glass tank to the orifice outlet. The transient temperature and velocity at the FLUENT/POLYFLOW interface are then mapped to a POLYFLOW sub-model as time-dependent boundary conditions. The glass gob free surface deformation process was modeled in the POLYFLOW sub-model.

Table 1 shows the computing time of modeling 3-D, transient glass gob formation with POLYFLOW and hybrid model. The POLYFLOW model has a cell count of 31,000 and the hybrid model has a cell count of 126,000. For isothermal cases, it takes 61 hours for POLYFLOW model to complete the calculation, and it takes 38 hours for the hybrid model to complete the calculation. By comparing the computing time with isothermal cases, we found that the hybrid model significantly improves the computation efficiency, by an average of a factor 2. Moreover, for non-isothermal cases, it takes 390 hours (more than 2 weeks), to complete the calculation via the POLYFLOW model. However, the computational time required by the hybrid model is only 3.63 days, for the same isothermal case.

Table 1: Computation time required by POLYFLOW and hybrid model

Computation Time	POLYFLOW (*Cell count: 31,000*)	Hybrid model (*Cell count: 126,000*)
Isothermal	61 hrs (2.54 days)	38 hrs (1.58 days)
Non-isothermal	390 hrs (16.27 days)	87 hrs (3.63 days)

For isothermal cases, the velocity contours and gob-shape predictions at 0.3s for the POLYFLOW model and hybrid model are compared in Fig. 11. In POLYFLOW model, the velocities of glass flow in the tank and in the gob were computed at the same time. In contrast, in the hybrid model, the velocity of glass flow in the tank was calculated by the FLUENT sub-model. The velocity of glass forming the gob was taken care of by POLYFLOW sub-model. The velocity magnitude patterns are similar, as illustrated in Fig. 11 (a) and (c). The maximum velocity magnitude predicted by hybrid model is 3.5 % higher than the one in the POLYFLOW model. The results from the POLYFLOW model and hybrid model show that when heat transfer is not taken into consideration, the gob has a cylindrical shape with semi-hemispherical ends. In this paper, we define the gob length as the vertical distance from the orifice outlet to the tip of the gob, as shown in Fig. 11 (b) and (d). In the POLYFLOW model, the gob length is 67.4 mm. In the hybrid model, the gob length is 71.4mm as shown, which is 5.6 % longer than its counterpart in the POLYFLOW model.

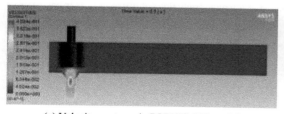

(a) Velocity contours in POLYFLOW model   (b) Gob shape in POLYFLOW model

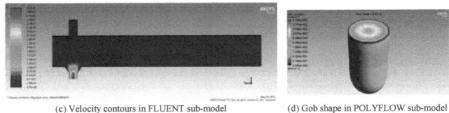

(c) Velocity contours in FLUENT sub-model   (d) Gob shape in POLYFLOW sub-model
Figure 11: 3-D glass gob formation at 0.3 s from isothermal model

For non-isothermal cases, the temperature contours and gob-shape predictions at 0.3 s by POLYFLOW model and hybrid model are shown in Fig. 12. Similar to the velocity contour comparisons for isothermal cases, the temperature contours have similar patterns with a maximum difference of temperature of 0.8 %, as plotted in Fig. 12 (a) and (c). As for the gob-shape predictions, both the POLYFLOW and hybrid models yield distorted cylindrical shapes due to temperature inhomogeneity in the gob as shown in Figs. 12 (b) and (d). The gob length predicted by the hybrid model is 69.4 mm. It is 4.7 % longer than the 66.1 mm length predicted by the POLYFLOW model.

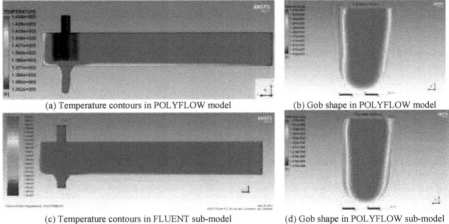

(a) Temperature contours in POLYFLOW model   (b) Gob shape in POLYFLOW model

(c) Temperature contours in FLUENT sub-model   (d) Gob shape in POLYFLOW sub-model
Figure 12: 3-D glass gob formation at 0.3 s from non-isothermal model

Figures 13 (a) and (b) summarize gob-length predictions at different time instances from the POLYFLOW and hybrid models for isothermal and non-isothermal cases, respectively. In general, the gob lengths increase linearly with time. The glass gob starts to form with an initial length of 12 mm at 0s. As time advances, the deviations between the results from POLYFLOW and the hybrid models increase. At 0.3s, the differences of gob-length are 5.6 % and 4.7 % for isothermal and non-isothermal cases, respectively.

The gob length is one of the factors to evaluate gob shape. The differences in gob-shape predictions between POLYFLOW and hybrid models might be caused by several reasons. First, the total cell count for POLYFLOW and hybrid models are different. The POLYFLOW model has 31,000 elements; the hybrid model has 126,000 elements. Different mesh density might also lead to differences in results. Secondly, the methods to treat moving objects are different. The POLYFLOW model employs MST to simulate the plunger motion and the hybrid model adopts the MDM in the FLUENT sub-model to simulate the moving plunger. One limitation of MST is once physical boundaries do not match mesh boundaries, the mass conservation equation cannot be satisfied in every element. By using MDM in FLUENT sub-models, no mass conservation issues exist. Thirdly, POLYFLOW is a Finite Element Method (FEM) based solver. The variables such as temperatures and velocities are computed at the nodes of elements. FLUENT is a CFD solver based on the Finite Volume Method (FVM) where temperatures and velocities are calculated at cell face centers, and then interpolated onto the adjacent nodes. This difference also causes a slight discrepancy in the result.

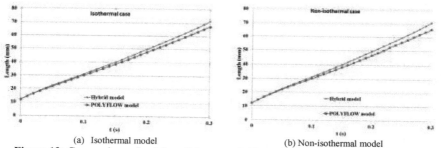

(a) Isothermal model   (b) Non-isothermal model

Figure 13: Comparison of gob length between POLYFLOW model and hybrid model

6. CONCLUSIONS

Two numerical approaches were developed to model the 3-D, transient, isothermal and non-isothermal glass gob forming process. In the first approach, ANSYS POLYFLOW was employed. The numerical results show the flow and thermal conditions of the molten glass in the feeder systems. The distorted cylindrical shape of a glass gob is caused by thermal inhomogeneity in the molten glass flow. In order to reduce the simulation time and improve the computational efficiency, an alternative method that combines the usage of ANSYS FLUENT and POLYFLOW was then introduced. Results show this hybrid model yields a significant reduction in computational time while maintaining reasonable accuracy.

7. FUTURE RESEARCH

Thermal radiation in the semi-transparent participating media of the molten glass is crucial to determine the overall heat transfer. Rosseland Approximation is the most widely used method in the glass industry for computing radiative heat transfer since it is the one of the computationally cheapest methods to implement. However, Rosseland Approximation requires the glass object should be optically thick; otherwise, this method is not accurate enough. In glass production, most processes are neither optically thick nor optically thin. That requires a thermal radiation model which could be applicable to all optical thickness material. Discrete ordinates

(DO) method is considered the most accurate radiative heat transfer model, and works for all optical thickness; however, the DO model is very computationally intensive.

Since the current study focuses on computational efficiency improvement by using the hybrid model, thermal radiation is not considered in this paper. In a future step, internal radiation will be modeled using the DO method. The influences of thermal radiation on the overall heat transfer and the free surface shape will be investigated later.

ACKNOWLEDGEMENTS

The authors would like to thank the GMIC for accepting this paper in the Glass Problems Conference, and the management of Owens-Illinois Inc. for providing the resources and time to complete this research. The authors also appreciate the help from Hossam Metwally and Lasya Reddy of ANSYS for numerous in-depth discussions on POLYFLOW modeling.

REFERENCES

1. Linden, B.J.V.D. (2002). *Radiative heat transfer in glass: The algebraic ray trace method.* Veenendaal, The Netherlands: Universal Press.
2. Brown, M. (2007). A review of research in numerical simulation for the glass-pressing process. *Proc. IMechE Part B J. Eng. Manufact.* , *221*(9), 1377–1386.
3. Hyre, M. R. (2002). Numerical simulation of glass forming and conditioning. *J. Am. Ceramic Soc., 85*(5), 1047-1056.
4. Hyre, M. R. (2004). Effect of mould to glass heat transfer on glass container forming. *Ceramic Trans., 141*, 271-279.
5. Liu, X. J., Hyre, M. R., Frost, G. S., & Austin, S. A. (2008). Numerical simulation of the heat transfer. *Proc. ASME, IMECE2008*-66675.
6. Hyre, M. & Rubin, Y. (2004). Modeling of gob and container forming. *Int. J. Forming Proc.,* *7*(4), 443-457.
7. Metwally, M. & Reddy, L. ANSYS Inc., Personal communication, 2013.
8. ANSYS POLYFLOW14.0 user's guide. ANSYS Inc. (2011)
9. ANSYS FLUENT 14.0 user's guide. ANSYS Inc. (2011)
10. Tooley, F.V. (1974). *The handbook of glass manufacture Vol II.* New York: Books for Industry, Inc..

MODELING OF HEAT TRANSFER AND GAS FLOWS IN GLASS FURNACE REGENERATORS

Oscar Verheijen, Andries Habraken, and Heike Gramberg
CelSian Glass & Solar B.V.
Eindhoven
The Netherlands

ABSTRACT

Improving energy efficiency and cost reduction in glass production are of key importance to maintain glass as cost-competitive product with environmental sound footprint. Regenerators of glass furnaces have a major impact both on energy efficiency in glass production and investment costs for new glass furnaces. The aim with designing of regenerators is to maximize heat recovery from the hot flue gases (and to preheat combustion air) while minimizing its volume (to limit purchasing expensive regenerator bricks) and ageing. In addition, the type of regenerator bricks applied as function of height in the regenerator (or better: as function of temperature in the regenerator), needs to be chosen such that it can chemically resist the attack/corrosion by the expected flue gas components at the prevailing temperature.

Optimal design of regenerators (in view of heat recovery, costs and lifetime) requires detailed 3D CFD simulations in order to determine the turbulent flows in the complete regenerator, the local temperatures of the gases and complex shaped regenerator bricks and the convective and radiative heat exchange between gases and checkers for both flue gas and air phase. This paper reports on results of detailed modeling of regenerators by CelSian's CFD model GTM-X. Next to 3D-temperature fields, the distribution of flue gas (and air) over the top (and bottom) cross-sectional checker layers, and the longitudinal and lateral flows further through the regenerator, depending on type of checkers and regenerator and port neck design is shown. In addition, critical areas for chemical fouling – either by sodium sulfate condensation or by attack of (especially the binder phases of) refractory material – is discussed as function of flue gas composition.

INTRODUCTION

Improving glass furnace energy efficiency is one of the key targets for glass companies to remain glass production a sustainable and cost-competitive industry. Average primary energy consumption of glass furnaces has decreased considerably since 1950's (see Figure 1, [1]) reaching an average level of about 4.8 GJ per ton produced container glass. Although a stabilization in improving glass furnace energy efficiency is observed in figure 1 since 2000, still significant energy savings can be realized for individual glass furnaces as is indicated by energy benchmarking data (see Figure 2): the best performing container glass furnace has a primary energy consumption level of about 3.3 GJ/ton (normalized to 50% cullet), whereas 90% of the container glass furnaces have a primary energy consumption of more than 4.2 GJ/ton.

One way of reducing energy consumption of regenerative glass furnaces is improving the heat recovery from flue gases by preheating combustion air. The theoretical maximum regenerator efficiency is in the order of 77%. However, practical values vary in the range of 60 – 65%. As each percent (absolute) increase in regenerator efficiency results in a reduction of energy consumption with about 0.9%, significant energy savings can be accomplished by improving the flue gas heat recovery behavior of regenerators. Besides reducing energy consumption of glass furnaces, improving the flue gas heat recovery in regenerators also might lead to a more compact regenerator design with lowered investment costs.

Optimization of the thermal performance of regenerators requires detailed modeling of the flow behavior of both combustion air and hot flue gas and the spatial temperature distribution in flue gas, combustion air and checker work.

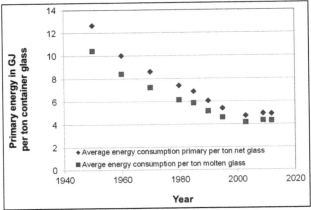

Figure 1. Average primary (triangle) and actual (square) energy consumption per ton of container glass in the Netherlands.

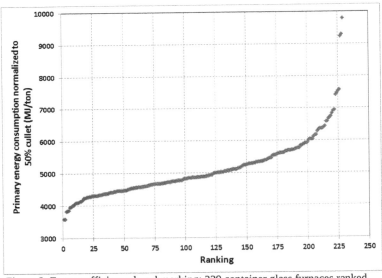

Figure 2. Energy efficiency benchmarking: 229 container glass furnaces ranked according to their primary energy consumption normalized to 50% cullet.

REGENERATOR MODELING

Detailed 3D modeling of glass furnace regenerators is required in order to analyze whether the regenerator volume is efficiently utilized and whether heat recovery from the hot flue gases by the checker work and the subsequent heat transfer towards the combustion air can be improved. Simulation of the flow pattern in regenerators reveals any presence of dead zones and/or flue gas and combustion air recirculation areas, which have a strong negative effect on regenerator efficiency. Next to its dependency on regenerator dimensions, flue gas heat recovery is determined by the types of checkers applied in the regenerator, the connection of the burner ports to the regenerator, and the presence and location of air infiltration. The uniformity of the combustion air and flue gas flow distribution on each cross-section of the regenerator, and thus the regenerator efficiency, is very much determined by the ability of lateral flow (cross-mixing) Lateral flow is determined by the shape and openness of the checker work and therefore, it is essential (see also [2]) that in regenerator modeling studies, the details of the different complex shaped checker types as indicated in Figure 3 are simulated.

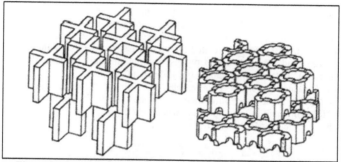

Figure 3. Flow patterns along different types of checkers (cruciform and pigeon holes) are simulated with regenerator modeling to accurately describe lateral flow of combustion air and flue gas

Detailed regenerator modeling not only provides valuable information on regenerator efficiency, and how to improve it, but also can be used to identify critical areas for fouling and clogging due to formation of condensates while cooling the flue gases. Additionally, the impact of air infiltration on regenerator and glass furnace energy efficiency can be quantified.

Until now, detailed modeling of complete regenerators has been hardly performed because of limited computational power, resulting in too long simulation time of the detailed flow and temperature fields in the large regenerators. Also, the complex shapes of checkers could not be included in Computational Fluid Dynamics (CFD) modeling, thus preventing adequate modeling of the 3D flow fields and local heat transfer rates from and to the checkers. To cope with the demands for regenerator modeling, CelSian Glass & Solar (further referred to as 'CelSian') extended the functionality of its simulation software code GTM-X. GTM-X is an industrially validated CFD program, developed and owned by CelSian, which is used for and by the international glass industry. CFD models define and solve conservation laws in order to calculate for instance fluid or gas velocities and temperatures. For this, the space of interest is divided into small cells (volume elements) that form the grid. In each cell, conservation of basic quantities such as mass, momentum, and energy is imposed; this leads to a very large number of coupled equations (typically in the order of millions) that are solved by the program.

However, currently GTM-X is able to use grids (Figure 4) that align with skew parts of the geometry (body-fitted), an essential feature in view of the complex shapes of most types of regenerator bricks. Moreover, cells can be divided in smaller sub-cells in those regions where this is needed, while cells can remain larger in regions where this can be afforded (local grid refinement). This reduces the total number of grid cells that is necessary to obtain an accurate solution. The so-called non-matching grid feature in GTM-X further allows to independently define a fine grid on specific locations which are not propagated into regions where it is not required. Also, GTM-X can use multiple processors to work on the same job (parallelization), and it efficiently deals with physical phenomena that are not the same in all regions of the calculation domain (multi-physics). These features significantly reduce the turn-around simulation time, as compared to conventional CFD programs. By these improvements, regenerator modeling becomes practically feasible.

With simulation of a double-pass regenerator (see Figure 5 for a schematic view) both heating of combustion air and cooling of the hot flue gases are simulated in parallel. Detailed heat transfer is described by applying advanced radiation models like Discrete Ordinate Method (DOM, [3]) to account for 3D radiation within the checkers, between checkers and refractory and between gases and checkers. For a so-called reference case (the 3D flow pattern and temperature distribution is modelled based on actual process settings and properties of the gas phases and refractory materials), the simulation model is validated against industrial process data, additional on-site process measurements, and photographs. In addition, thermochemical simulations are performed to predict condensate formation dependent of the measured industrial flue gas composition. The composition and location where condensates are formed is validated with chemical analyses of deposits collected from the regenerator.

Figure 4. Detailed view of grid used for cruciforms (left) and chimney blocks (right).

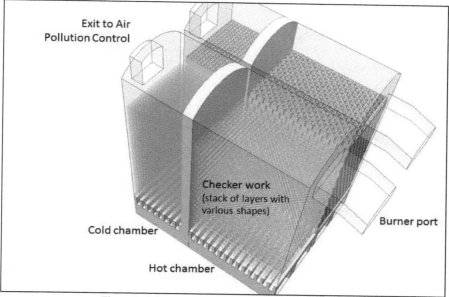

Figure 5. Schematic view of double-pass regenerator

CONDENSATION OF SODIUM SULFATE

During furnace lifetime, the specific energy consumption of glass furnaces generally increases between 5 to more than 15%. Next to the increase in air filtration and chemical attack of refractory materials by aggressive flue gas species, formation of condensates from the flue gas during cooling is a key parameter affecting regenerator (and thus overall furnace) efficiency. The composition of the condensates depends on flue gas composition and temperature. In regenerators of soda-lime-silica glass furnaces, the condensates are mainly sodium sulfates. Sodium sulfate is formed by a reaction of sodium hydroxide and sulfur oxides in the flue gas according to:

$$2\ NaOH\ (g) + SO_2\ (g) + \tfrac{1}{2}\ O_2\ (g) \quad \rightarrow Na_2SO_4\ (s,l) + H_2O\ (g)$$

Sodium components mainly evaporate from the molten glass by reaction of the sodium oxide in the melt with water vapor in the furnace atmosphere, leading to the formation of volatile sodium hydroxides. Sulfur dioxide is formed by decomposition of sulfates both in the batch blanket and the glass melt. The condensation of sodium sulfate depends on temperature and concentration of the relevant species in the flue gas. Accurate calculation of the flue gas and checker temperatures enables prediction of the critical areas in the regenerator that are prone to clogging by means of Na_2SO_4 condensate formation. Figure 6 shows the sodium sulfate condensation temperature as function of oxygen content of the flue gas for three sets of vapor pressures of NaOH and SO_2. The sodium sulfate condensation temperature decreases with decreasing oxygen content, SO_2 content and NaOH content.

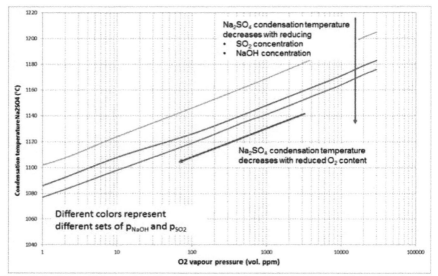

Figure 6. Sodium sulfate condensation temperature as function of oxygen content of the flue gas for three sets of vapor pressures of NaOH and SO_2.

CONCLUSIONS

This paper describes the need for detailed and fast simulation models for glass furnace regenerators to enable optimization of heat recovery from hot flue gases. CelSian's GTM-X simulation software is very well suited for these cpu-intensive calculations due to its specific functionalities that significantly reduce calculation time.

Combination of detailed modeling with industrial measurements enables a complete analysis of the regenerator performance, including actual evaluation of flow pattern and temperature distribution due to air leakages. In addition, flue gas analysis, thermochemical calculations, knowhow on refractory material and chemical analysis of condensates helps in clarifying fouling and clogging issues. Thus, the complete analysis enables the identification of critical zones that are sensitive for regenerator corrosion and provides recommendations to minimize or prevent corrosion issues.

REFERENCES
[1]   R.G.C. Beerkens, 'Trends in Glass Production – Innovation or Slowdown', Lecture of the Otto Schott Memorial Medal awardee at the 1st Joint Meeting of DGG – ACerS GOMD, Aachen (Germany), May 25-30, 2014.
[2]   David Lechevalier, Olivier Citti, Sebastien Bourdonnais, 'New refractory cruciform for improved energy efficiency of regenerative glass furnace', ICG 2010, Salvador (Brazil), 20-25 September 2010.
[3]   A.M. Lankhorst, L. Thielen, P.J.P.M. Simons, A.F.J.A. Habraken, Proper modeling of radiative heat transfer in clear glass melts', 73rd Conference on Glass Problems, Cincinnati, Ohio (USA), October 1-3, 2012, pp 249-258.

ENERGY ANALYSIS FOR PREHEATING AND MODELING OF HEAT TRANSFER FROM FLUE GAS TO A GRANULE

Liming Shi, Udaya Vempati, and Sutapa Bhaduri
Owens-Illinois, Inc.
Perrysburg, OH 43551, USA

ABSTRACT

To reduce energy consumption of an oxy-fired glass furnace, energy loss via the flue gas and through the furnace walls should be minimized since they account for approximately 40 % of the energy usage. One way to minimize flue gas energy loss is to pre-heat the raw batch materials and/or cullet. Energy analysis was conducted to evaluate the maximum amount of energy recoverable and that required for the pre-heating. The maximum pre-heating temperature was calculated under conditions of constant and reduced rate of natural gas usage. Since the flue gas temperature from an oxy-fired furnace is on the order of 1350 °C, the maximum temperature for batch pre-heating that could be potentially employed was above 500 °C. However, handling loose batch at such high temperatures is likely to be physically difficult to accomplish reliably. On the other hand, batch in an agglomerated form, such as granules, may be pre-heated easily. Analysis of the heat transfer from flue gas to a single granule was investigated first through computational fluid dynamics (CFD) modeling. Parameters studied included the average diameter and thermal conductivity of the granule, the inlet flue gas temperature, and the flue gas velocity and composition. The data was used to evaluate the time needed to preheat a single representative batch granule to a given target temperature under various heating conditions. In addition, the time-dependent temperature and velocity distributions for the modeled geometry were determined. The results show that granule diameter and gas velocity both have a significant impact on the rate of granule heating.

INTRODUCTION

One of the advantages of an oxy-fuel fired furnace is its low NO_x emission. The oxy-fuel furnace is more promising if its efficiency of energy utilization can be further improved. For example, energy loss via the flue gases and through the furnace walls account for approximately 40 % of the energy usage. Currently, flue gases leave oxy-fuel fired furnaces at about 1350 °C or above, but the inherent thermal energy is typically not utilized because of the danger associated with preheating the incoming oxygen or natural gas fuel. One means of reducing the energy loss associated with the hot flue gas is to utilize it in a Thermo-Chemical-Recuperator (TCR) [1] to convert methane into synthesis gas. The estimated energy savings for a float glass oxy-fuel furnace are on the order of 20 %. However, additional investigation is needed to directly evaluate the impact of TCR on the glass production process, and to understand the corrosive effects of the hot flue gases on the active TCR materials.

Another option to improve energy utilization is to preheat raw batch and/or cullet. The impact of pre-heating on glass production is well understood. When using commercial pre-heaters [2], the flue gases need to be diluted to ~600 °C prior to contacting the raw batch due to possible corrosion effects on the duct material. There are also other issues associated with direct contact heat transfer between loose batch and hot flue gases, such as batch segregation, carryover of fine particles and handling of large volume flue gases. In addition, some loose batch compositions may undergo local incipient melting and/or release of significant amounts of moisture, both of which can cause batch clumping and plugging in the pre-heater prior to entering the furnace.

Batch in granule format may withstand higher temperature flue gases than loose batch since the material is already "pre-clumped" or agglomerated to a controlled size and shape. Because the raw batch constituents (sand, soda ash, limestone, etc) are in more intimate contact with one another within the granule or briquette, mixing distances are reduced which can lead to lower melting temperature and better homogenization of the melt in the furnace. In addition, the use of granulated batch will largely mitigate segregation. Prior to developing and designing a pre-heater that could employ batch granules, an energy analysis was conducted to estimate the maximum preheating temperature that could be theoretically achieved.

Since granules are agglomerated batch material, their diameters are significantly larger the size of the constituent grains. In order to determine the time needed to heat a granule to a target temperature and achieve temperature uniformity within the granule, a Computational Fluid Dynamics (CFD) approach was used to model the heat transfer process. Parameters investigated in the analysis include the average diameter and average thermal conductivity of the granule, the inlet gas temperature, and the flue gas velocity and composition.

This paper addresses the method and results from energy analysis and CFD analysis of heat transfer from flue gas to a granule.

ENERGY ANALYSIS FOR PREHEATING
Method
The program used for energy analysis was developed in-house for container glass furnaces. It is based on mass balance and energy balance. It can be used to calculate energy input, energy loss through the furnace walls, and energy carried by flue gases based on input information. Raw materials include sand, soda ash, limestone, and minor constituents. The assumptions used in energy analysis are as follows:

1. Natural gas is the fuel for the oxy-fuel furnace.
2. Any reactions that occur between batch components during pre-heating process are not considered.
3. Heat loss is not considered during the dilution of flue gases.

Given conditions include a pull rate of 250 ton/day and a cullet percentage in the batch of 50 %.

Results
The energy of the flue gases relative to 25°C is shown in Figure 1 as a function of exit temperature of flue gases. These energies are the theoretical maximum energy for pre-heating. In this calculation, flue gas composition (i.e. the ratio of CO_2, H_2O, etc.) is assumed to be invariant of exit temperature. As shown in Figure 1, energy of flue gases is decreased by about 15.6 % as the exit temperature of flue gases is reduced by 200°C. That is, the energy of flue gases depends on the exit temperature of flue gases.

In reality, flue gases with temperature above 1300°C are most likely diluted by air before preheating. Thus the actual energy available for pre-heating is affected by the exit temperature of flue gases and the extent of dilution. The recoverable energy is the difference between the total energy in the diluted flue gases and the total energy of the flue gases leaving the pre-heating system. Figure 2 shows the recoverable energy as a function of diluted flue gas temperature. It is assumed that ambient air is used for dilution without thermal loss due to quenching. The exit temperature of flue gases from the furnace is assumed to be 1350°C. The solid black line represents the theoretical maximum energy available in flue gas relative to 25°C.

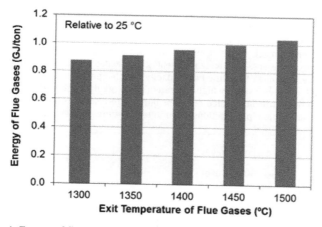

Figure 1. Energy of flue gases relative to 25°C versus exit temperature of flue gases

The other two curves shown in Figure 2 are based on flue gases leaving the pre-heating system at 100°C and 200°C, respectively. As expected, the recoverable energy decreases as the exit temperature of the flue gas increases; however, it should be noted that an exit temperature of 200°C is more likely to be achieved in a practical pre-heater system. Over a single curve (i.e. conditions in which the flue gas exits at the same exit temperature from the pre-heater system), the recoverable energy reduces with decreasing temperature of diluted flue gases. This is because the total mass flow rate of flue gases leaving the pre-heating system is higher at a lower dilution temperature, which contains more energy. For example, the total mass flow rate for the diluted flue gases at 800°C is almost 2.1 times of that at 1350°C.

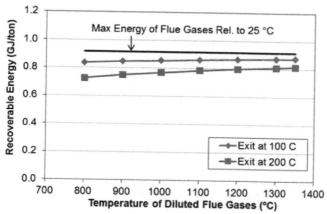

Figure 2. Recoverable energy versus temperature of flue gases after dilution

Figure 3 shows the theoretical energy required to pre-heat the raw batch granules and the cullet in a 50-50 batch mixture. Note that the maximum cullet pre-heating temperature is assumed to be 450°C because container glass cullet will become soft and sticky above this temperature. For either batch or cullet, the theoretical pre-heating energy increases as the pre-heating temperature increases. Relatively more energy is needed at higher temperatures for batch due to the non-linear change of specific heat capacity with temperature. For the same pre-heating temperature, batch requires a slightly higher theoretical energy than cullet. It should be noted that the amount of energy pertinent to batch reactions during pre-heating is not considered. Based on preliminary experimentation, pre-reactions occur at temperatures greater than 700°C.

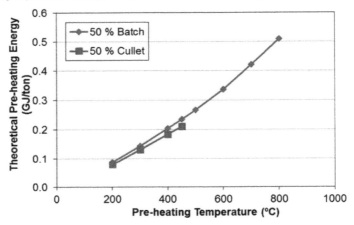

Figure 3. Theoretical Pre-heating Energy versus Pre-heating Temperature for Batch and Cullet

The maximum pre-heating temperature can be calculated by comparing the theoretically required pre-heating energy for batch or cullet and the recoverable energy from flue gases (i.e. the information in Figures 2 and 3, respectively). It was assumed that thermal efficiency for heat transfer from flue gases to batch or cullet typically varies from 70% to 85%. Figure 4 shows the maximum pre-heating temperature for batch as a function of the diluted flue gas temperature over the above heat transfer efficiency range. One assumption used in Figure 4 is to keep the rate of natural gas usage constant. For batch compositions consisting of 50% cullet (50% raw batch or "50% batch"), it was again assumed that the maximum cullet pre-heating temperature is 450°C. The maximum pre-heating temperature for batch, consisting of either 50% or 100% raw batch (noted in the Figure as "50% raw batch" or "100% raw batch," respectively), decreases as the temperature of diluted flue gases reduces because of less recoverable energy from the flue gases. The maximum pre-heating temperature is higher at higher thermal efficiency. When thermal efficiency decreases from 85% to 70%, the difference in the maximum pre-heating temperature is around 140°C for 50 % batch and 80°C for 100% batch.

The maximum pre-heating temperatures for 100% batch are lower than those for 50% batch under the same heating conditions. This is because more energy is required by 100% raw batch than 50% raw batch and 50% cullet. There is a limit on the maximum pre-heating temperature of cullet (450°C), while no limit on the maximum pre-heating temperature of raw batch. It can be seen that the maximum pre-heating temperatures are in the range of 480°C to

770°C. When the calculated pre-heating temperatures are above 700°C, the actual pre-heating temperatures that can be practically achieved will be lower because of endothermic batch reactions that occur above 700°C. According to the experiments of Wu and Cooper [3], the sticking characteristics of loose batch will also sharply increase within a shorter time period at higher temperatures. For example, they noted that loose soda-lime-silicate batch starts to become sticky upon thermally soaking at 770°C for five hours. They also reported a "no sticking" condition at temperatures below 745°C. Experiments are required to evaluate analogous sticking behavior of agglomerated batch granules.

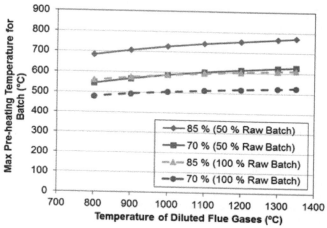

Figure 4. Maximum pre-heating temperature for batch versus temperature of diluted flue gases

When using pre-heated batch and cullet, the rate of natural gas usage is reduced, compared to the base case of batch and cullet that is not pre-heated, because additional energy is carried into the furnace by pre-heating. Consequently, the reduction in natual gas usage leads to reduced flue gas amount and, thus, recoverable energy. An iterative calculation is needed to establish the maximum pre-heating tempreature achievable when the natural gas usage is reduced. This was calculated by balancing the energy recoverable under reduced rate of natual gas usage (properly accounting for thermal efficiency) with the energy required for pre-heating batch and cullet. Figure 5 shows the results of the maximum pre-heating temperature under constant and reduced rate of natural gas usage for 50 % raw batch. The pre-heating temperature varies in the range of 444°C to 627°C under reduced rate of natural gas usage. In comparison with the maximum pre-heating temperatures under constant rate of natural gas usage, those under reduced rate are about 134°C less at the thermal efficiency of 85%, and 114°C less at the thermal efficiency of 70%. In Figure 5, the pre-heat temperature for 50% cullet is set at 450°C. The only exception is the maximum pre-heating temperature for batch and cullet is about 444°C (below 450°C) when the temperature of diluted flue gases is 800°C.

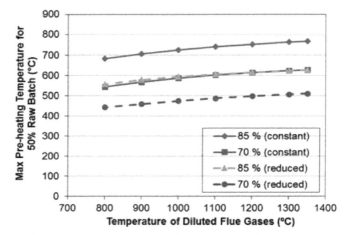

Figure 5. Maximum pre-heating temperature for 50 % raw batch versus temperature of diluted flue gases at constant and reduced rate of natural gas usage

Based on the assumed 70% to 85% thermal efficiency of the pre-heating system, an energy savings in the range of 12% to 19% is estimated. The energies associated with various inherent reactions, such as water dissociation and carbonate calcination, need to be measured and incorporated to improve these calculations. The purpose of the above analysis is to present an overall comparison between energy available and energy required for pre-heating and estimate the maximum pre-heating temperature for batch and cullet.

CFD ANALYSIS OF HEAT TRANSFER FROM FLUE GAS TO A GRANULE
In this section, heat transfer from the flue gas to a single granule was modeled using CFD. The following results are useful in designing a pre-heater.

Description of Modeled Geometry
The modeled geometry for heat transfer analysis is shown in Figure 6. It is 0.06 m in length, 0.02 m in depth, and 0.02 m in width. The position of a single granule is fixed in the flow field. The overall flow field is sufficiently large to minimize effects from boundary on the flow field near the granule. Flue gases with a given temperature enter the duct from the inlet, pass across the granule, and leave at the outlet.
Method
The commercial software ANSYS FLUENT was used for CFD analysis. Unsteady flow was modeled to obtain the variation of granule temperature with time. The k-ω with shear-stress transport turbulence model was used. The modeled flue gases included four gas species, referring to CO_2, H_2O, N_2, and O_2. The boundary condition was velocity-inlet at the inlet, pressure outlet at the outlet, and symmetry for the four sides. Second-order upwind scheme[4] was used for the momentum, species, and energy equations.

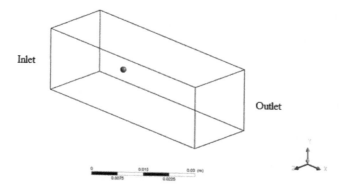

Figure 6. Modeled geometry for granule heat transfer analysis

Parameters for Sensitivity Analysis
Thermal Conductivity of Granules
     Literature information for the measured thermal conductivity of a loose glass powder batch is very limited. The reported results from measurements made by Kröger and Eligehausen on the thermal conductivity of powder batch from 20°C to 1250°C [5,6] are plotted in Figure 7. The thermal conductivity slowly increases with temperature from 20 °C to 750 °C, but increases very sharply with temperature from 750°C to 1250°C. The sharp increase is due to the appearance of a molten phase in the batch at these higher temperatures. Based on the energy analysis described above, the maximum pre-heating temperature for batch is 770°C. At this temperature and below, the thermal conductivity of glass batch is low because it consists of a solid phase only. Therefore, an average default value of 0.455 W/m-K was used for the thermal conductivity of granule.

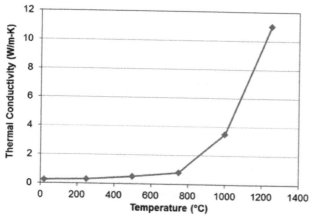

Figure 7. Thermal conductivity versus temperature for powder batch [5,6]

Specific Heat Capacity of Granules

Sharp and Ginther [7] found that the specific heat capacity of glass varies from 968 J/kg-K to 1188 J/kg-K as temperature increases from 300°C to 1000°C. Pilon et al. used a value of 1100 J/kg-K for batch in their study [8]. Since the variation of specific heat capacity with temperature is small, a constant value of 1100J/kg-K was also used in this study.

Density of Granules

As mentioned previously, the granule is assumed to consist primarily of sand, soda ash, and limestone in a typical container glass composition, the theoretical density of which is 2300kg/m^3 to 2500kg/m^3. Since a batch granule contains void space, a density of 2012kg/m^3 was employed.

Listed in Table 1 are the default values used in the CFD models. For initial calculations, the diameter of granules remained between 1.19mm and 2.38mm, the average of which was assumed as the default. The default flue gas composition is representative of that in a typical oxy-fuel fired furnace. It usually contains a high content of CO_2 and H_2O. Due to limitations on the expected materials of construction for a pre-heating system, the default temperature of the incoming flue gas was assumed to be 800°C.

Table 1. Default values used in the CFD models

Parameters	Default Values	
Diameter of granule	1.785 mm	
Thermal conductivity of granule	0.455 W/m-K	
Inlet temperature of flue gases	800 °C	
Inlet velocity of flue gases	1.0 m/s	
Composition of flue gases (Oxy-fuel exhaust gas)	CO_2	40.31 vol%
	H_2O	51.73 vol %
	O_2	2.59 vol%
	N_2	5.37 vol%

RESULTS

Parameters for the analysis included the diameter and thermal conductivity of the granule, and the inlet flue gas temperature, velocity, and composition. Flue gas velocity and temperature profiles and the temperature distribution within the granule were determined from the simulations. In this analysis, the assumption is that flue gases only pass through a single granule. The effect of other particles on the flow and temperature of flue gases and the particle/particle interaction are not considered.

The profile of gas speed

The velocity profile at the center plane (XY) of the modeled geometry after 10s of pre-heating is shown in Figure 8. All the parameters were set at their default values, as listed in Table 1, and a no-slip boundary condition was applied to the surface of granule. While the total simulation time was 20s, most of the change in the gas speed profile occurred within the first 3s. It can be seen that the gas flow (left to right in the Figure) is deflected before reaching the solid granule and moves around it. Immediately in front of the granule (the left hand side), there is a small region of very low velocity gas. Behind the granule (the right hand side), is a larger area of low velocity gas, with gas velocity gradually increasing along the flow direction. That is, it takes longer distance for the gas flow to re-merge behind the granule than for the flow to separate in front.

Figure 8. Gas speed profile at the center plane of the modeled geometry at the end of 10s

The temperature history of the granule

The initial temperature of granule was 25°C. As hot flue gas passes around the granule, the granule's temperature increases with time. Radiation was ignored in the simulation because the inlet temperature of the flue gas was not very high. The temperature profile at the center plane of the modeled geometry is shown in Figure 9 at three different times. The default values shown in Table 1 were used in the model. It can be seen that the granule temperature is on average below 154°C at 0.5s, ~600°C at 5s, and ~700°C at 10s. That is, granule temperature quickly increases within the first 5 s of pre-heating, but increases more slowly thereafter. The gas temperature behind granule is lower than the bulk because it just transfers heat to the granule. The length of lower gas temperature zone behind the granule is longer when the gas gives more energy to the granule. This relatively lower temperature region behind the granule becomes smaller with increasing time.

To better understand the variation in granule temperature during pre-heating, Figure 10 shows the average temperature of the whole granule and those at five point locations expressed as a ratio with the starting temperature (T_0) as a function of time. Points 1 to 5 represent the temperature at: the front surface of the granule, a quarter of a diameter after Point 1, the center of the granule, three quarters of a diameter after Point 1, and the back surface of the granule, respectively, which can be seen in Figure 10. Temperature differences at each of these locations are apparent over the first 8s of pre-heating and negligible after that. In comparison with the time dependent average temperatures, temperatures at Point 1 are higher because this point directly faces the hot flue gases. The temperature at Point 2 is very close to the average temperature for each measure of elapsed time. Temperatures at Points 3 to 5 are slightly lower than the average temperature, but they are very close to each other. Heat is conducted within granule and this transfer depends on the thermal conductivity of granule.

For each location within the granule, the temperature increases with time exponentially. The initial increase in temperature is fast because of the large temperature difference between the granule and flue gases. As the temperature difference becomes smaller, the increase in granule temperature declines. For example, it takes about 5.2s, 6.2s, 9.2s, and 12.1s for the average temperature ratio (T/T_0) to reach 0.75, 0.80, 0.90, and 0.95, respectively.

The effect of granule diameter

A series of simulations were conducted with the diameter of the granule set at 0.595 mm, 1.19mm, 1.785mm, and 2.38mm, respectively. Other parameters given in Table 1 were kept constant. Figure 11 shows the ratio of average temperature of granule to the inlet flue gas temperature (T/T_0) as a function of time at four granule diameters. At each diameter, the average temperature increases with time exponentially. The initial slope of the curve is much steeper for the smaller size granules. It takes about 1.9s, 4.6s, 9.2s, and 14.9s for the size of 0.595mm, 1.19mm, 1.785mm, and 2.38mm granule to reach $T/T_0=0.9$, respectively. Therefore, a longer time is required for the larger size granule to reach the temperature of the flue gases. This is probably because the thermal mass of the granule is reduced and the surface area per unit volume is increased at smaller size.

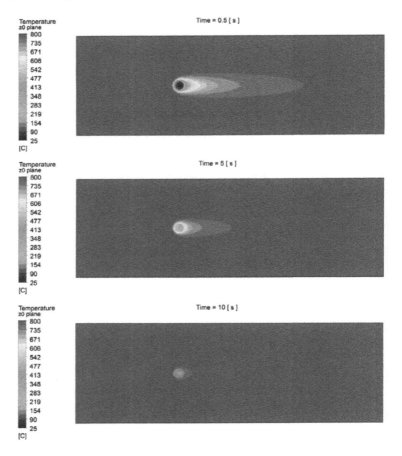

Figure 9. Temperature profile at the center plane of the modeled geometry at three different times

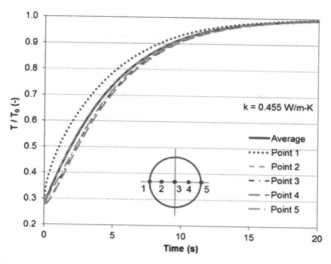

Figure 10. Temperature tatio versus time for the average and those at five locations within a granule when thermal conductivity of granule was 0.455 W/m-K.

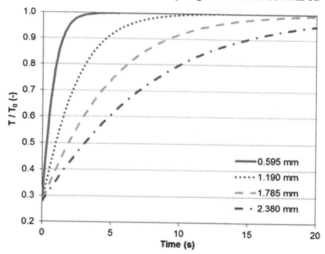

Figure 11. Average temperature ratio versus time at four granule diameters

The effect of the thermal conductivity of the granule
    Figure 12 shows the ratio of average temperature of granule to the inlet gases temperature (T/T_0) as a function of time at four granule thermal conductivities, 0.25W/m-K, 0.455W/m-K, 2.0W/m-K, and 5.0W/m-K respectively, which were chosen as possible thermal conductivities

near the value Kröger and Eligehausen reported [6]. The average granule temperature is relatively smaller for the lower thermal conductivity, but this difference diminishes as the thermal conductivity rises above 2.0W/m-K. Within the chosen range thermal conductivity has a relative small effect on the average temperature of the granule during pre-heating. This indicates convective heat transfer from flue gases to the granule is the dominant mechanism controlling the temperature of granule.

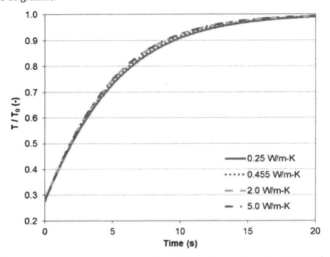

Figure 12. Average temperature ratio versus time at four thermal conductivities of granule

To evaluate the effect of thermal conductivity on the temperature profile inside of the granule, the ratio of temperature difference between Point 1 and Point 5 to the temperature of inlet gases was calculated. Figure 13 shows this ratio $((T_1-T_5)/T_0)$ as a function of time at four thermal conductivities. At each thermal conductivity of granule, the temperature difference initially increases with time, reaches a peak, and decreases afterwards. The time corresponding to the peak becomes longer with decreasing thermal conductivity. At least for the first 8s, lower thermal conductivities lead to larger temperature difference $((T_1-T_5)/T_0 >0.05)$ because of weaker heat conduction within the granule.

The effect of inlet temperature of the flue gases

The inlet temperature of flue gas was set at 500°C, 600°C, 800°C, 900°C respectively for a series for simulations conducted to examine the effect of the incoming flue gas temperature on granule pre-heating. Several commercial batch/cullet pre-heaters use flue gases at temperature of ~500°C. The upper limit on the inlet temperature of flue gases often depends on the material used in constructing the pre-heater. Figure 14 shows the average temperature ratio (T/T_0) as a function of time at four inlet temperatures of flue gases. In general, the temperature ratio is affected slightly by the inlet temperature of gases. During the initial 5s, temperature ratio slightly increases as inlet temperature decreases, but this is reversed after 6s. The gas temperature affects gas properties such as density, viscosity, and thermal conductivity, which impacts the convective heat transfer coefficient from the gases to the granule.

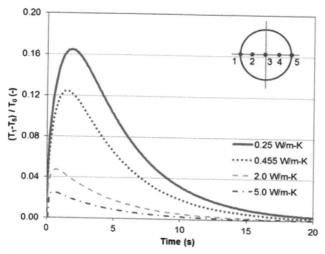

Figure 13. Ratio of temperature difference to the inlet temperature of flue gases versus time at four thermal conductivities of granule

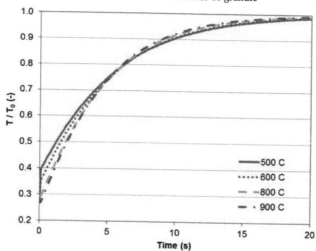

Figure 14. Average temperature ratio versus time at four inlet temperatures of flue gases

The effect of the inlet velocity of the flue gases

The inlet velocity of flue gases was set at 0.25m/s, 1m/s, 2.5m/s, and 10m/s respectively for a series of calculations conducted to examine the effect of the speed of the incoming flue gas on granule pre-heating. A wide range of inlet velocities was considered because this variable is highly dependent on pre-heater design. The average temperature ratio (T/T_0) as a function of time at four inlet velocities is shown in Figure 15. The effect of inlet velocity on granule temperature is apparent. The temperature ratio increases with increasing inlet velocity very quickly from 0.25m/s to 2.5m/s, but more moderately from 2.5m/s to 10m/s. That is, the increase in temperature does not change linearly with the inlet velocity. The required time to reach 90% of the initial gas temperature ($T/T_0 = 0.9$) reduces from 12.9s to 4.3s as the inlet velocity increases from 0.25m/s to 10m/s. It was found that the convective heat transfer coefficient increases significantly at higher inlet velocities of the flue gas, which in turn leads to a faster rate of heat transfer. When inlet velocity is held constant, the convective heat transfer coefficient decreases with time as the temperature difference between the granule and the flue gases diminishes.

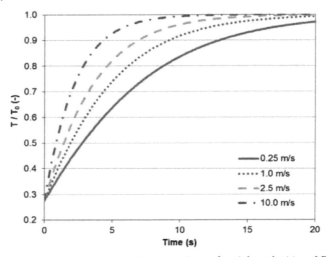

Figure 15. Average temperature ratio versus time at four inlet velocities of flue gases

The effect of flue gas composition

    The exhaust gases leaving an oxy-fuel fired furnace usually have high temperatures, around 1350°C to 1450°C. Most likely they will be diluted by air to a lower temperature before subsequent pre-heating or other applications or treatments. The composition of the flue gas changes after dilution. To determine the effect of flue gas composition on granule heat transfer, simulations were conducted at two conditions. The corresponding compositions of the two variants are given in Table 2. One represents a typical oxygen-fuel furnace exhaust gas and the other an exhaust gas that has been diluted by air to reach a temperature of 800°C. The amount of CO_2 and H_2O accounts for about 92% in volume before dilution and 44% in volume after dilution. Figure 16 compares the average temperature ratio (T/T_0) versus time at the two-modeled conditions. The effect of gas composition on the granule temperature is negligible. This indicates that the variation in the overall properties of the two-modeled gases is small.

Table 2. Composition of flue gases for the two modeled conditions

Gas Composition	Oxy-fuel Exhaust Gases (vol %)	Oxy-fuel Exhaust Gases with Dilution to 800 °C (vol %)
CO_2	40.31	19.20
H_2O	51.73	24.65
O_2	2.59	12.23
N_2	5.37	43.92

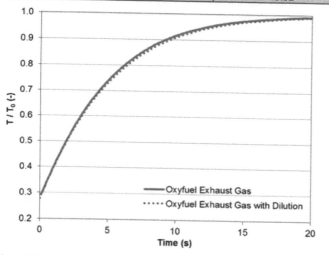

Figure 16. Average temperature ratio versus time at two modeled conditions

Estimation of the height of a pre-heater

The above energy and CFD analysis were used to estimate the height of a pre-heater with counter-current flow. In this arrangement, batch granules move downward and are directly contacted the upward flowing flue gases. This is similar to the advanced raining bed concept [9]. According to CFD analysis, the desired residence time was about 3s to 5s as a first estimate for delivering nearly uniformly pre-heated granules. Under this condition, the height of the pre-heater is estimated to be 38m to 64m for 1.785mm granules. By using granules with a diameter of 0.595m, the pre-heater height can be reduced to 12m to 20m to achieve the same amount of residence time. Therefore, the height of the pre-heater is sensitive to the size of granule, and becomes shorter with smaller size granules. Modifying the internal structure of the pre-heater (e.g. by using baffles) may further extend the residence time of the granules due to dispersion.

CONCLUSIONS

An in-house developed energy analysis program was used to determine the amount of energy available from flue gases and energy required by batch/cullet for pre-heating. Under the assumed conditions, the calculated maximum pre-heating temperature for batch consisting of 50%

cullet and 50% raw batch was around 480°C to 770°C at constant rate of natural gas usage and 444°C to 627°C at reduced rate of natural gas usage.

Heat transfer from the flue gases to a single granule was modeled using a CFD approach. Temperature and velocity distribution in the modeled geometry, including the granule itself, can be predicted. The time required to pre-heat the granule to a given target temperature could be readily obtained. The results of sensitivity analysis show that granule diameter and the inlet velocity of the flue gases both have a significant effect on the average temperature of the granule, whereas the inlet temperature of the flue gases and the thermal conductivity of the granule each display a slight effect on the average temperature ratio, and the composition of the flue gas exhibits a negligible effect. The thermal conductivity of the granule affects heat transfer within the granule itself. The granule pre-heating process can be sped up by increasing the velocity of the incoming flue gases, which causes enhanced convective heat transfer from the flue gas to the granule, and/or by reducing the diameter of the granules.

In the future, the amount of heat associated with the calcination reactions should be included in both the energy and CFD analysis to provide greater accuracy in the simulations. The physical behavior of the granule, such as sticking, under various temperatures should also be measured. Both analyses are useful in defining an initial pre-heater concept. In addition, a more detailed CFD analysis can be used to accelerate the pre-heater design process and better understand flow and heat transfer within various conceptual designs.

ACKNOWLEDGEMENT

We would like to thank Dr. Scott Weil who led us through this study. We also would like to thank the R&D of Owens-Illinois for providing financial support for this project.

REFERENCES
1. van H. Limpt, A. Suarez-Barcena, E. de Wit, "Thermo-Chemical-Recuperator Technology: A Big Step towards Energy Efficient Glass Melting," DGG-GOMD Conference, May 28, (2014), Aachen, Germany.
2. R. Beerkens, "Energy Saving Options for Glass Furnaces & Recovery of Heat from their Flue Gases and Experiences with batch & Cullet Pre-heaters Applied in the Glass Industry, " 69th Conference on Glass Problems: Ceramic Engineering and Science Proceeding, 30, 1, 143-162, (2009).
3. Y. Wu, A.R. Cooper, "Batch and Cullet Preheating Can Save Energy," Glass Industry, 10-13, (1992).
4. ANSYS Inc., "ANSYS Fluent User's Guide," (2013).
5. NCNG, TNO, Glass Technology Course Textbook, Part 1, 198-200, (2012).
6. C. Kröger, H. Eligehausen, "Über das Wärmeleitvermogen des einschmelzenden Glasgemenges," Glastech Ber, 32, 9, 362-373, (1959).
7. D.E. Sharp, L.B. Ginther, "Effect of composition and temperature on the specific heat of glass", Journal of the American Ceramic Society, 34, 260-271, (1951).
8. L. Pilon, G. Zhao, R. Viskanta, "Three-Dimensional Flow and Thermal Structures in Glass Melting Furnaces. Part II: Effect of Batch and Bubbles," Glass Science and Technology, 75, 3, 115-124, (2006).
9. R.W. Breault, A. Litka, A.W. McClaine, R.P. Chamberland, D.T. McNeil, T. Wilsoncroft, "An Integrated Cullet/Batch Preheater System for Oxygen-Fuel Fired Glass Furnaces," American Flame Research Committee, September 30-October 2, (1996), Baltimore, Maryland.

LABORATORY FACILITIES FOR SIMULATION OF ESSENTIAL PROCESS STEPS IN INDUSTRIAL GLASS FURNACES

Mathi Rongen, Mathieu Hubert, Penny Marson, Stef Lessmann, and Oscar Verheijen

CelSian Glass and Solar, Zwaanstraat 1
5651 CA Eindhoven
The Netherlands.
Ronxing Bei
RHI AG
Wienerbergstrasse 9
A-1100 Vienna, Austria

ABSTRACT
Glass melting is often viewed as a black box process: raw materials and energy are used as input, and the glass product results. The complex processes which occur in the glass melting tank are of vital importance for the resulting amount and quality of the glass obtained and the conditions should be such that the occurrence of unwanted side effects such as refractory corrosion and high emission of waste gases are minimized. Improving the glass making process requires complex modeling and experimental observation of the detailed sub processes. CelSian Glass & Solar has developed and uses several unique dedicated laboratory set ups and methods to simulate in detail the major process steps which take place in industrial glass melting furnaces. The facilities are tailor-made for studying industrial glass melting processes, simulating industrial conditions, furnace atmospheres, batch compositions, use of industrial refractories and heating curves. In this paper, two of the laboratory set-ups are described in detail and examples are given which demonstrate their application.
- Testing the flue gas attack of regenerator refractory materials to compare the corrosion resistance of the refractories.
- Use of high temperature melting observation and evolved gas analysis to study foaming and ways to combat this.

INTRODUCTION

The complex processes which occur in the glass melting tank are of vital importance for the resulting amount and quality of the glass obtained, and the conditions should be maintained such that the occurrence of unwanted side effects such as carry over, evaporation, foaming, refractory corrosion and high emission of waste gases, as well as energy consumption are kept to a minimum. Improving the glass melting process requires complex modeling and experimental observation [1].

However experiments in industrial glass furnaces are risky and extremely difficult to carry out because of the harsh conditions in the furnaces, high temperatures and limited positions of access to the glass melt. Also in industrial glass production, no changes can be made which may disturb the glass melting process and have potential detrimental effect on the product quality.

Therefore CelSian Glass & Solar has developed and uses several dedicated laboratory set ups and methods to simulate in detail the major process steps which take place in industrial glass melting furnaces. The detail process steps include:
- Melting behaviour of raw materials
- Removal of gases (bubbles & dissolved gases) from the melt (fining)

- Re-absorption of gas bubbles during cooling of the melt (re-fining)
- Foaming (reducing energy efficiency of the furnace and attacking sidewall refractories)
- Evaporation from the melt (causing dust emissions, refractory corrosion)
- Carry-over of batch components
- Refractory corrosion by the atmosphere in regenerators and melting tank

Dedicated laboratory equipment is available to study (amongst others) the following important process steps:

- Gas evolution during melting & fining (EGA)
- High temperature observation of melting, foaming & fining
- Gases dissolved in the glass
- Evaporation from glass melts
- Regenerator and crown refractory corrosion
- Carry-over and decrepitation
- Redox state during melting
- Batch heating - heat transfer into the batch blanket
- Batch free time studies
- Bubble free time studies

In this paper two experimental set ups are highlighted: combined High Temperature Observation of Melting and Evolved Gas Analysis and the study of refractory corrosion by flue gas.

EXPERIMENTAL STUDIES ON REFRACTORY CORROSION
DESCRIPTION OF THE EXPERIMENTAL EQUIPMENT

A sketch of the principle of the set up used at CelSian to investigate refractory corrosion by simulated flue gases is shown in Figure 1. A gas burner is used to create a combustion atmosphere. The natural gas from the burner can be burnt with pure oxygen, with air or a mixture of these two. The fuel, oxygen and air supply are precisely controlled with mass flow controllers and in this way the concentration of O_2 and CO in the combustion gases can be controlled to simulate the conditions of the atmosphere of an industrial glass furnace. In this way, it is therefore possible to create an oxidizing or reducing atmosphere. In this example a sodium hydroxide (NaOH) solution is injected into the flame. This reacts with the also injected SO_2 gas and forms Na_2SO_4 when the gas flows into the (cooler) electrically heated tube furnace. The simulated flue gas flows over the refractory samples, which are placed on heat resistant metal wires that are wound around a ceramic tube so that all sides of the refractory are exposed to the gas. The tube furnace is temperature controlled enabling a well-defined temperature gradient between 970°C and 770°C. The different refractory types are exposed to the gas in one run and subjected to this temperature gradient. The refractory materials corrosion as function of the temperature to which they are exposed can thus be studied. The concentrations of O_2, CO_2, SO_2 and CO are measured with analyzers. Sodium and sulfur content are measured by using wet chemical extractions methods. After the experiment, each of the refractory material samples is analyzed in a scanning electron microscope (SEM).

Set-up:

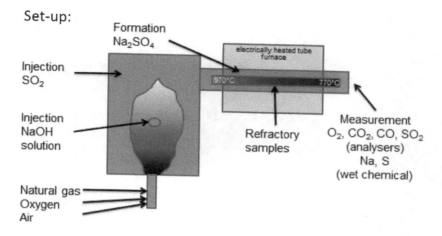

Figure 1. Sketch of refractory corrosion set-up

A photograph of the assembly that constitutes the experimental set up is shown in Figure 2. Figure 3 shows the refractory material samples in the tube furnace.

Figure 2. Photograph showing the refractory corrosion set-up

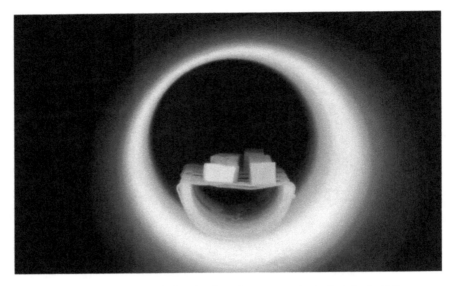

Figure 3. Photograph showing the refractory samples in the tube furnace

EXAMPLE OF STUDY OF CORROSION OF REGENERATOR REFRACTORIES

Refractory corrosion trials are essential to understand reactions occurring inside regenerators to ultimately select the best refractory types and to design regenerators with a longer campaign time.

Magnesium oxide refractories are used in the middle section of regenerator stacks. In this example two different MgO refractory materials were tested under oxidizing conditions, one with calcium silicate as binder and the second material with a magnesium silicate (forsterite) binder. The MgO refractory material samples (size approx. 4 x 1 x 1cm) were exposed to flue gases containing NaOH and SO_2 for a period of 10 days. After 10 days each refractory sample had gained in weight, as shown in Figure 4.

SEM analyses of MgO refractory samples containing the calcium silicate binder showed the presence of Na_2SO_4 and free SO_3. At higher exposure temperatures, more weight increase was observed and more Na_2SO_4 was detected. The calcium silicate binder was corroded by SO_3. Calcium sulfate was formed as shown in the SEM image in Figure 5.

A SEM image of a forsterite bonded (2) MgO (1) refractory sample is shown in Figure 6. A lower weight gain was observed for the forsterite bonded MgO. Slight corrosion was observed, as the forsterite protects the attack of MgO grains by SO_3.

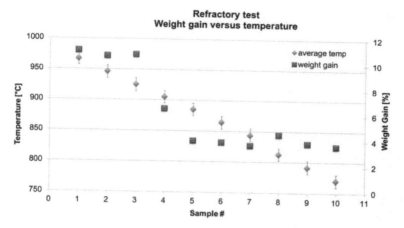

Figure 4. Refractory test weight gain versus temperature

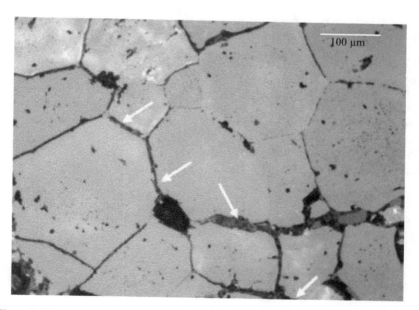

Figure 5. SEM image showing corrosion of the calcium silicate binder between the MgO grains (denoted by arrows).

Figure 6. SEM image showing less corrosion of the MgO grains in forsterite bonded MgO

EXPLANATION
As noted in [2] and confirmed by thermodynamic calculations with FactSage, in the temperature range 750 °C to 1000 °C calcium silicate reacts with SO_2 and oxygen to form calcium sulphate $CaSO_4$ according to reaction (1) below:

$$2Ca_2SiO_{4(s)} + SO_{2(g)} + 1/2O_{2(g)} \rightarrow CaSO_{4(s)} + Ca_3Si_2O_{7(s)} \qquad \text{Reaction (1)}$$

$$Mg_2SiO_{4(s)} + SO_{2(g)} + 1/2O_{2(g)} \rightarrow MgSO_{4(s,l)} + MgSiO_{3(s)} \qquad \text{Reaction (2)}$$

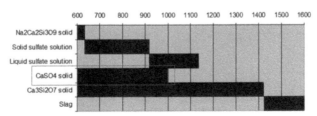

Figure 7. Thermodynamic calculation (FactSage) showing the formation of calcium sulphate in temperature range between 600 and 1000°C

A similar reaction (2) of magnesium silicate (forsterite) with SO_2 and oxygen to form magnesium sulphate $MgSO_4$ only occurs up to about 625°C as shown in Figure 8.

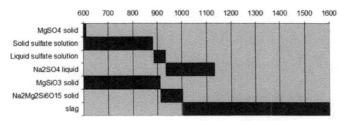

Figure 8. Thermodynamic calculation (FactSage) showing that formation of magnesium sulphate only occurs below 625°C

Therefore the use of forsterite as binder protects against severe corrosion in the temperature range 700-1000°C [3]. The upper operating temperature is limited due to liquid slag phase formation starting around 1000°C.

CONCLUSIONS ON EXPERIMENTAL STUDY OF REFRACTORY CORROSION
The refractory testing experimental set up used at CelSian can be applied to check the corrosion resistance of refractory materials under oxidizing or reducing conditions, different flue gas compositions, and temperatures. The set up can also be used to determine the admissible CO concentrations (when applying near stoichiometric combustion) to minimize energy consumption without causing fouling or clogging of the regenerators. The particular example shown here demonstrates how the choice of the binder used for MgO refractories is critical for corrosion resistance.

HIGH TEMPERATURE MELTING OBSERVATION AND EVOLVED GAS ANALYSIS
DESCRIPTION OF THE EXPERIMENTAL EQUIPMENT:
A schematic drawing of the experimental set-up developed for the Evolved Gas Analysis (EGA) and simultaneous glass melting observation is shown in Figure 9. A constant carrier gas flow (gas mixtures applied to simulate the industrial furnace melting atmosphere) set by a mass flow controller is led into a tall vitreous silica crucible containing the batch. The top part of the vitreous silica extends above the furnace roof and is hermetically sealed with a cooled lid. Two ceramic tubes pass through the lid and function as inlet and outlet tubes for the gases. Gases that are released from the batch/glass mix with the carrier gas and are removed from the crucible atmosphere through the exit pipe. Subsequently, the gas mixture is led by heated pipes to the Fourier Transform Infrared Spectrometer (FTIR), MKS Model 2030 for flue gas composition analyses.
The volume fractions of several gases are measured by the FTIR, in particular the evolution of CO, CO_2 and SO_2. Gases that leave the FTIR are passed through a cooler, to remove the water vapor (the amount of water vapor in the melting atmosphere can be controlled to simulate industrial glass melting atmospheres such as melting in oxy-fuel conditions, with approx. 55 vol% water in the combustion chamber). The resulting dry gas is passed on to zirconia oxygen analyzers, which can measure concentrations of oxygen at different levels, from some volume % to the ppm scale.
The experimental set up is also equipped with a high resolution CCD camera. Pictures of the batch/glass melt in the vitreous silica crucible are taken at regular intervals. These images can be used for, for example, detailed observation of the reactions taking place, onset of melting phase formation, measurement of foam amount, and determination of the fining onset

temperature. A film can be prepared from the recorded images. Comparing films from different experiments shows clearly the effect of, for example, furnace atmosphere on the foaming tendency.

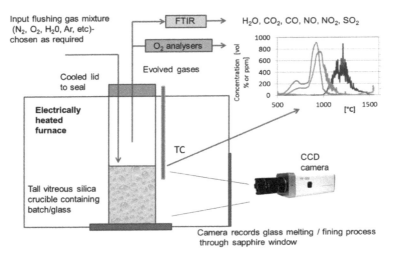

Figure 9. Schematic representation of the CelSian's Evolved Gas Analysis and Melting Observation Set-up (EGA-HTMOS)

Photographs of the assembly that constitutes the experimental set up are shown in Figures 10 and 11.

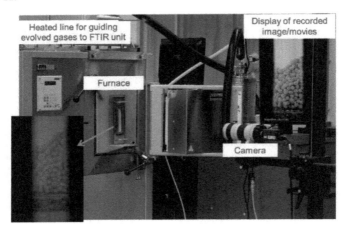

Figure 10. CelSian's evolved gas analysis and high temperature melting observation apparatus

Figure 11. Evolved gas analysis equipment containing FTIR and oxygen analyzers

EXAMPLE OF STUDY OF PREVENTION AND BREAKDOWN OF FOAM IN INDUSTRIAL FURNACES PRODUCING BOROSILICATE GLASS

In this paper we show experimental results from investigations on the parameters, which influence the tendency of a glass batch/melt to foam and the amount of foaming. The parameters which can influence the foaming tendency and which are included are the viscosity, surface tension, raw material grain sizes, raw materials used.

Practical methods to prevent foaming or encourage breakdown of foam, which can be examined are: the effect of minor additions to the batch, the effect of (temporary) changes in the furnace atmosphere, and spraying or dripping of solutions to destabilize any foam.

Experiments can be carried out on different types of glasses (soda-lime-silicates, borosilicates, special glasses). High temperature observation enabled examination of the foam layer, its thickness and structure. The foam and glass level can be determined from the high temperature images as a function of time, as shown in Figure 12.

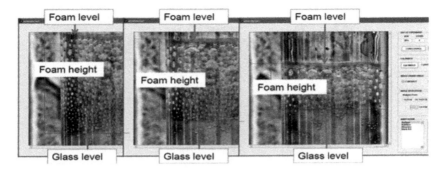

Figure 12. Measurement of foam layer thickness from high temperature images

In the example below, the effect of adding cokes to a borosilicate glass batch is demonstrated by comparing the gas releases of the 2 glasses as a function of temperature. For these experiments 150g of each batch was melted in a transparent sealed vitreous silica crucible according to a set temperature profile. The temperature and concentration of the gases released (during melting, foaming and fining) are plotted as a function of time in Figure 13.

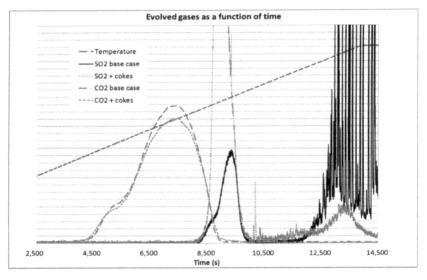

Figure 13. Evolved gases as a function of time for a borosilicate glass from batch with (+ cokes) and without cokes (base case)

Figure 13 demonstrates that addition of a small amount of cokes to the batch encourages earlier (lower temperature) release of SO_2, as seen also for various glass melts [5, 6]. In this case, when gas evolution peaks are combined with HTMOS images, the dramatic effect of the addition of a small amount of cokes on reduction of the foaming behaviour at high temperature can be observed for this particular batch. The melting, fining and foaming behaviours are characteristic to each batch composition melted in given conditions (melting atmosphere, temperatures involved). The effect of all these parameters can be evaluated with CelSian's EGA-HTMOS equipment.

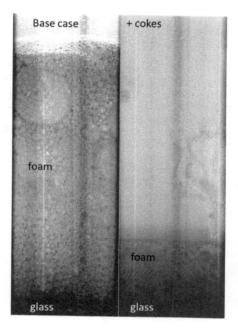

Figure 14. High temperature image showing the difference in foaming behaviour of the base case and + cokes glasses

CONCLUSIONS HIGH TEMPERATURE MELTING OBSERVATION AND EVOLVED GAS ANALYSIS

Evolved Gas Analysis & High Temperature in-Situ Observation during batch-to-melt conversion provides a valuable method for many investigations. These can include investigating ways to melt-in faster by applying a certain batch grain size distribution, use of melting aids, using pellets, or other batch pre-treatment methods. Typical melting-in reactions being endothermic, their occurrence at lower temperatures can lead to considerable energy savings.

New batch compositions can be studied to see the effects on foaming, melting-in & fining behaviour and to optimize a batch for fining. The use of alternative raw materials can be explored, as can the effect of furnace atmosphere on fining onset (temperature) or foaming. The equipment is well-suited to determine foam decay rates, or the fining onset temperature (for example to be used in CFD models to study fining behaviour in industrial tanks with certain glass melt flow patterns and temperature levels). Investigating of the effect of batch routing/selective batches or batch pre-treatment on melting & fining rates is another interesting application of these procedures.

REFERENCES

[1]   R.G.C Beerkens. Simulation of essential process steps in industrial glass melting. 23rd International Congress on Glass, Prague 2013.

[2]   Routschka et al. Studies on the behaviour of magnesia, spinel and forsterite refractory bricks under simulated service conditions in the middle regions of oil-fired glass furnace regenerators. Glastech. Ber. Glass Sci. Technol. 63 (1990), no.3, pp. 61-68.

[3]   Routschka et al. Studies on the behaviour of magnesia, spinel and forsterite refractory bricks under simulated service conditions in the middle regions of oil-fired glass furnace regenerators. Glastech. Ber. Glass Sci. Technol. 63 (1990), no.4, pp. 87-92.

[4]   C. Wöhmeyer; G. Routschka Corrosion resistance of binding and matrix phases in basic checker bricks for the middle regions of glass furnace regenerators. Glastech. Ber. Glass Sci. Technol. 63 (1990), no.5, pp.118-126.

[5]   J. Collignon; M. Rongen; R. Beerkens: *Gas release during melting and fining of sulfur containing glasses.* Glass Technol.: Eur. J. Glass Sci. Technol. A, (2010), **51,** no. 3, pp.123-129

[6]   R. G. C. Beerkens: *Fining of Glass Melts: What we Know about Fining Processes Today.* Proceedings of the 69th Conference on Glass Problems, 4.-5. November 2008. Columbus OH, Am. Ceram. Soc. J. Wiley & Sons Inc. (2009) pp. 13-28

HEAT TRANSFER IN GLASS QUENCHING FOR GLASS TEMPERING

Carlos J. Garciamoreno, David A. Everest and Arvind Atreya*
Department of Mechanical Engineering
University of Michigan, Ann Arbor, MI 48109

ABSTRACT

This paper presents the results of an experimental study of heat transfer characteristics in single-phase and two-phase stagnation point flows pertinent to quenching of glass in the tempering process. Two-phase flows were generated by injecting water mist into the stagnation flow air far upstream of the nozzle exit. This resulted in a temporal and spatially invariant size distribution of the droplets that were carried toward the hot test plate by the air flow. PIV measurements were made at the nozzle exit to determine the magnitude and uniformity of air velocity profile in both single-phase and two-phase flows. The two-phase flows were also characterized by measurements of drop size distribution and number density using images of droplets resulting from laser induced fluorescence. The ratio of nozzle–to–plate distance and the nozzle diameter was maintained at 0.5 throughout the experiments. Steady state experiments were performed for plate heat fluxes ranging from 10 to 50kW/m², and Reynolds numbers ranging from 2,000 to 122,000 and water/air mass flow ratios up to 4.75%.

Single-phase flow results indicate that the Reynolds number dependence of the Nusselt number is ~ $Re^{0.68}$. Two-phase flow results show a maximum heat transfer enhancement of 26% for water/air mass flow ratio of 4.75%. It was visually determined that for plate temperatures above 200°C and for the drop size distribution tested, the water droplets do not impinge on the plate surface. Therefore, the heat transfer enhancement was attributed to the evaporation of water droplets within the thermal boundary layer. This is an important condition to prevent spatially non-uniform quenching and the resulting shattering of glass. Transient characteristics of single-phase and two-phase flows were also analyzed and compared. By changing the water/air mass flow ratio, the cooling curve for a two-phase flow can be adjusted to meet the requirements of the industrial process.

INTRODUCTION

Single-phase air jet impingement cooling is widely used in many industrial processes, such as glass tempering, annealing of metals, cooling of turbine blades, cooling of electronic equipment, etc. To increase the cooling rate, the air flow velocity must be increased, which adds to the energy required by the blowers. An alternate method of increasing the heat transfer rate in cooling processes is needed to reduce the air flow velocity and hence the energy consumption by the blowers. Mist cooling has been used to cool steel in the limit of film boiling, and work by Sozbir [1] indicates that the mist cooling can be useful in the tempering of glass.

The addition of water mist into the main stream air flow will enhance the cooling capacity by utilizing the latent heat of evaporation of the liquid phase**. This method provides an additional way of controlling the cooling rate by controlling the amount of water addition. It has a potential to decrease the energy consumption by the blowers and/or increase the productivity by increasing the cooling rate. However, depending on the material, water droplet impingement on the surface of the material being cooled may be detrimental. Consequently, the conditions required for avoiding water droplet impingement on the surface and the magnitude of cooling enhancement possible without impingement must be quantified to evaluate the usefulness of the technique. The quantification of cooling enhancement by the addition of mist is the subject of the present study. Classical stagnation-point flow geometry is used and the heat transfer rate, velocity and drop size distributions are measured in two-phase flows and compared with single-phase flow.

Experimental and theoretical studies of mist cooling were conducted over three decades ago [2, 3]. These studies showed that the heat transfer performance of a single-phase flow can be significantly improved by adding mist into the primary air flow. A number of two-phase flow studies have been made for a variety of configurations [4, 5]. Results show that latent heat of evaporation within the boundary layer is important and the formation of a thin liquid film over the surface will greatly enhance heat transfer rates. A constraint of the current study was to avoid the deposition of drops on the heated plate, therefore only evaporation of the liquid in the boundary layer will be considered.

Two-phase jet impingement cooling for the case where an atomizing nozzle is aimed directly at the surface has also been studied. Ohkubo and Nishio [6] studied the transient mist cooling for thermal tempering of glass plates at 690°C and showed that the heat transfer could be enhanced by water addition. The heat transfer coefficient was observed to be proportional to $V^{0.6}$, where V is the volumetric-droplet-flow-rate per unit surface area in m^3/m^2s. Further, they showed that the surface material property can have an effect on the heat transfer rate. However, drop sizes were not measured and the approach velocities were large (>20 m/s) leading to droplet impingement. Study by Liu [7] indicates that both the liquid mass flow and the mean drop diameter are important in the limit of low mass flux range and wall superheat temperatures less than 100°C. For surface temperatures lower than 100°C, Graham and Ramadhyani [8] reported that the heat transfer is primarily due to the evaporation of the liquid film at the surface. In the current study, the drop velocity field has been measured, the drop size distribution quantified and super heat temperatures are greater than 100°C to avoid drop impingement on the plate.

For the case where the mist is generated by an atomizing nozzle and injected into the primary air flow, at high wall temperatures Yao and Choi [9] showed that the heat transfer is dependent on the water mass flux. Li et al [10] also found a heat transfer enhancement due to mist added to steam flow. The largest effect was noticed in the stagnation region. The current study involves small Weber numbers ($We = \rho D V^2/\sigma \sim 10$) at low to moderate water mass flow rates on a plate with a uniform temperature profile in a flow that simulates the stagnation region.

The goal of the present study is to investigate the heat transfer rate in the regime of surface temperatures and droplet sizes where the water droplets evaporate within the boundary layer and do not impinge on the surface. None of the above cited studies have attempted to avoid droplet impingement during cooling. This is necessary because several industrial cooling applications are sensitive to surface wetting. The magnitude of heat transfer enhancement in the dry surface limit is measured and the transition region where the droplets begin to impinge on the plate is determined. To properly quantify the flow parameters, the drop size distribution and the velocity field is measured using the laser based Wide Field Particle Sizing and Velocimetry (WFPSV) technique which simultaneously measures the particle size and Velocity.

EXPERIMENTAL SETUP

Flow System
A schematic diagram of the flow system is shown in Fig. 1. The blower determines the main stream flow velocity in the wind tunnel, and it is controlled by an autotransformer. The range of flow velocities studied was from 0.39 to 19.05 m/s and the corresponding Reynolds numbers based on the exit diameter were 2,500 < Re < 122,000. The highest velocity, 19.05 m/s, is similar to the velocities used in industrial cooling systems. A velocity of 3.9 m/s, approximately 5 times lower than the highest velocity, was chosen for two-phase flow studies.

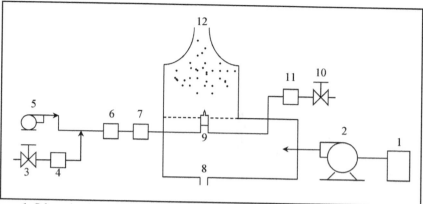

Figure 1: Schematic of flow system. 1) Autotransformer, 2) Blower, 3) Water shut-off valve, 4) Water filter, 5) Dye pump, 6) Flowmeter, 7) Pressure gauge, 8) Drain, 9) Nozzle, 10) Air shut-off valve, 11) Pressure gauge, and 12) Nozzle exit.

The flow system is configured such that the blower forces room air through the 0.61m (2') straight section of the tunnel before changing the direction of flow upward. It then passes through two 100-mesh screens, which make the velocity profile practically uniform. As shown, a humidifying nozzle discharges water droplets into the flow after the screens. This nozzle is operated by supplying water at 140KPa (20psi) and compressed air at 480 to 690KPa (70 to 100psi). The water flow rate (up to 1.9 ℓ/min), compressed air pressure and water pressure can be controlled to change the drop size distribution and the spray angle. The spray pattern is a hollow cone, whose angle can be varied from 100° to 147°. The highest momentum droplets are directed toward the walls of the tunnel, minimizing the disruption introduced in the main air flow and allowing the biggest droplets to hit the wall and fall down to the water collection section. A hole at the bottom of the tunnel is used to drain and measure this excess water. It is then subtracted from the total water flow rate to yield the water carried by the air flow. Only small water droplets are incorporated into the flow and a fine mist is obtained. After the water mist is introduced into the flow and allowed to mix, the wind tunnel cross section is reduced from a square 0.305m×0.305m (1'×1') cross section to a circular cross section 0.102m (4") in diameter using a smoothly varying contoured nozzle to reduce the boundary layer growth. After this reduction, the flow passes through a final 0.102m (4") diameter, 0.25m (10") long section before exiting.

To control the water mist introduced into the air flow, the water pressure is monitored by a 0–414KPa (0–60psig) pressure gauge, and controlled by a precision needle valve. The water flow rate is monitored by an Omega FL-100 flow meter, with a flow range from 0.1 to 780ml/min. The air pressure is monitored by a 0–690KPa (0–100psig) pressure gage and the air flow rate is monitored by an Omega FL-73 flow meter, with a flow range of 28–254ℓ/min (1–9ft^3/min). A Cole-Parmer Masterflex pump was used to inject fluorescent dye solution into the water flow before it is directed to the humidifying nozzle.

Hot Plate

A schematic of the heater assembly used to heat the test plate is shown in Fig. 2. A solid tower structure supports the hot-plate assembly, allowing it to translate vertically and rotate horizontally. The size of the heated square brass test plate (7.62×7.62×0.32cm) is chosen to be smaller than the exit diameter of the air nozzle. Thus, the edges of the air jet hit the unheated portion of the heater assembly allowing a more uniform velocity profile flow to be directed at the heated plate.

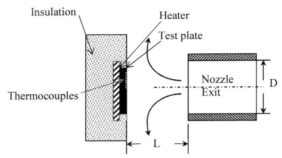

Figure 2: Schematic diagram of the furnace with heated test plate.

The heat source is a ring heater, sandwiched between the insulation and the brass test plate, with a maximum operating temperature of 650°C and a nominal power output of 500W at 115V. Its dimensions are: ID=4.28cm, OD=10.08cm and a thickness of 0.79cm. The power output is controlled by a variac and measured with an Extech digital multi-meter.

Insulation with a minimum thickness of 1 inch was placed around the heater and the plate to minimize heat losses. The insulation used was Kaowool 3000HT ceramic board, which can withstand temperatures up to 3000°C and has a thermal conductivity of 1.834W/m°K (1.06BTU·in/hr·ft^2·°F) at 554K (538°F). All components are enclosed in a stainless steel box with dimensions of 20.32×30.48×5.08 cm (8″×12″×2″). Holders were designed to keep the test plate flush with the cover to avoid protrusions that may cause flow disturbances. The two holders were located on corners diagonally across from each other and were separated from the brass test plate by 0.635-cm thick ceramic insulation to minimize heat losses. The ceramic insulation used is ZIRCAR Alumina Insulation Type AL-30, which has a thermal conductivity of 0.2076 W/m°K (0.12BTU·in/hr·ft^2·°F) at 525°C.

To measure the temperature eight K-type thermocouples 0.076mm (0.003in) in diameter were installed in the brass test plate. A narrow channel was carved in the surface of the test plate with enough space to install the bare wires and fill up the remaining space with cement. The cement used was an Omega Omegabond OB-400, with a thermal conductivity of 19.03W/m°K (11Btu-in/ft^2-hr-°F) and maximum service temperature of 1427°C. Although this cement works as an electrical insulation between the thermocouple wires and the test plate to prevent imprecise measurements, its high thermal conductivity ensures that the thermocouple measures the actual temperature of the metal plate.

The separation between the successive thermocouples was 5.4mm. The first thermocouple is located at the stagnation point, in the center of the plate, and the next seven are located toward the edge in a straight line as indicated in Figure 2. From the welded joint of each thermocouple, the positive wire runs toward one side of the plate and the negative in the opposite direction. The wires run between the stainless steel cover and the insulation and exit the heater assembly from the top where they are connected to the data acquisition system through a set of 0.51mm (0.02in) K-type extension wires. The Hewlett-Packard 3497A data acquisition unit used has a maximum reading rate of 100 readings per second at 3.5 significant digits. For the present measurements, the data acquisition system was set to take ~54 readings per second with a 3.5-

digit precision. Nine measurements (8 channels + time) were taken at a rate of approximately 6 Hz. This data frequency was higher than any changes in the plate temperature. As seen in Figure 3, the measured radial temperature profiles across the test plate demonstrate that the temperature was uniform for both conditions – with and without water addition.

WFPSV System

Wide Field Particle Sizing and Velocimetry is a new technique developed by Everest and Atreya [11] and Putorti et al. [12] to photographically determine the size and spatial distributions of particles in a large flow field. Here, laser induced fluorescence from a tracer dye in the droplets is imaged by a photographic camera at high resolution and analyzed to determine its size

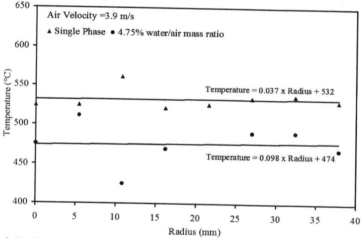

Figure 3: Radial temperature profile across the test plate for two conditions: 1) with 3.9 m/s nozzle exit air velocity and 2) same air velocity and 4.75% water/air mass ratio addition.

and location. While droplet velocities can also be simultaneously measured by this technique, here it is only used to measure only the droplet size.

In the present experiments, a pulsed Nd:YAG laser was used to provide 210 mJ/pulse of 532-nm beam that was expended into a light sheet using a 75-mm fused silica plano-concave cylindrical lens which was placed 26.7 cm from the center of the image. Rhodamine-based dye solution was injected into the water flow at a rate of 8 gm/gal. This dye was excited by the laser beam resulting in broad band fluorescence in the red wavelengths. A Schott OG 550 filter was used to separate the 532nm scattered light from the fluorescence signal from the droplets.

A 35mm Nikon N8008S film camera was used to image the fluorescent droplets. The camera was synchronized with the laser using an external switch and a solid state relay connected to the laser synchronizer. The camera shutter was set to $1/15^{th}$ of a second to allow only one laser pulse to be captured during each exposure and minimizing background noise.

A Nikon 80-200mm lens with a bellows extension of 83mm and five extension tubes (27, 14, 31, 21 and 13mm) provided a magnification of 2.17. The aperture of the camera was set to 4 and ASA 800 film was used to provide the best signal/noise ratio. The film negatives were scanned by a Polaroid scanner with a resolution of 4000 pixels per inch, resulting in an object resolution of 2.93µm per pixel. Drop sizes from these images were determined by using ImageProPlus 4.1 software.

A Sobel filter was used to calculate the gradient of the drop intensities. A threshold limit was used to define the outer diameter of the drop. The outer diameter of each drop was calculated by averaging the lengths of diameters measured at 2-degree intervals. The threshold on the gradient image will cause holes to form in the low gradient regions inside the drops. This area was used to calculate an equivalent inner diameter and the average of the two diameters was taken as the drop diameter. Images were filtered by aspect ratio and by color to remove particles that were out of the laser sheet plane or resulted from scratches caused by film processing and scanning. Visual inspection of the images indicated that the aspect ratio should be 2.5 of less and the particle color should be red. No particles less than 17µm were observed during visual inspection.

PIV System

The PIV system consists of two Nd:YAG lasers and a PIV camera to obtain two images of the flow with a known time separation. The particle image displacements are measured at many points in the image by the software to produce the flow field. The PIV system used for this study can be divided in the following subsystems: seeding, laser system, synchronizer and image capture camera.

Both single-phase air velocity and two-phase droplet velocities were measured. Magnesium oxide seeding particles were used to measure the single-phase flow velocities. The seeder volume flow rate was measured by a flow meter with a range of 0-28ℓ/min (0-60ft^3/hr) and the pressure was measured by a pressure gauge with a range of 0-690KPa (0-100psig). The MgO particles are atomized by compressed air and sent to a distribution chamber inside the wind tunnel downstream of the mesh screens. The percentage of compressed air introduced relatively to the total air flow was 0.76, 1.56 and 7.64% for exit velocities of 7.62, 3.9 and 0.39m/s, respectively. In the two-phase flow, the water drops were used as the seeding particles.

The laser system used consists of two collinear pulsed Nd:YAG lasers. The output of both lasers is frequency doubled to produce short duration (~5ns) high energy (150 to 210mJ/pulse) pulses of light at 532nm. This type of laser is well suited for PIV applications since the pulse energy is high enough to illuminate micron size particles in air, while the pulse duration is short enough to freeze the motion of even supersonic flows. The laser sheet produced by the laser and the associated optics was about 5mm thick. Since this thickness is large relative to the particles size and its low power edges can cause inconsistencies in the PIV measurements, it was reduced to 2mm using two sliding metal sheets and the center of the light sheet was used.

A TSI Model 610030 Synchronizer was used as an external controller for the laser and the image capture camera. This computer-controlled synchronizer provides the timing and sequencing of events for PIV measurements. The most important parameter controlled by the synchronizer is the time separation between the two laser pulses. The time separation was adjusted from 2ms for exit velocities of 0.39 m/s down to 100µsec for exit velocities of 7.62 m/s.

The image capture camera used for the PIV experiments was TSI PIVCAM 10-30. This camera has an asynchronous double exposure mode that allows a pair of images to be captured in less than 300µsec after an external trigger signal. The lens aperture was set to 4 for most of the experiments, allowing the maximum amount of light to be gathered. The images are analyzed by the TSI Insight software using a cross correlation routine and a 1.3 cm interrogation region.

EXPERIMENTAL PROCEDURE AND DATA ANALYSIS

Steady State Experiments

For these experiments, the heater input, air velocity and water addition were held constant until the thermocouples had achieved steady state. After the steady state temperature was recorded, the heat input was changed and the system was allowed to move to a new steady state. The heat input range for these experiments was 100-550 Watts, resulting in plate temperatures between 50 and 550°C. The accuracy of the input power measurements was ±0.75% for voltage and ±1% for current, which results in a power input error of ±1.75%.

For each steady state heat flux, seven temperatures were measured for every condition of velocity and water/air mass flow ratio investigated to ensure that a reliable trend could be determined. The radial profiles of temperature at 500°C without water addition and with 4.75% water/air mass flow ratio are shown in Figure 3. The temperature profiles are flat or slightly increasing toward the edges and have errors of 2.2% and 5.2% for no water addition and 4.75% water/air mass flow ratio respectively. The steady state temperature was chosen to be that of the center zone of the plate, 2.54 cm in diameter. The three thermocouples used to record the temperatures in this zone exhibited a uniform temperature profile.

A repeatability study was performed in which the experiments were done by approaching steady state both from higher and lower temperatures. Steady state was determined when the temperature change was less than 1°C for 300 seconds. The error caused by imposing this limit for defining the steady state was +3% if the steady state is reached from a lower temperature state and -3% if reached from a higher temperature state.

An important part of these experiments was determining if droplets were impacting the hot plate surface and if a water film was developing on the test plate. A digital video camera Sony DCR-TRV30 was used to record the condition of the hot test plate surface illuminated by white light. Two-phase flows with water/air mass flow ratios of 1.64 and 2.81% were recorded. For these water/air mass flow ratios, water droplets were visually observed to not impact the plate surface for temperatures higher than 200°C, and above this temperature the heat transfer curves for two-phase flows are similar to those of single-phase flows. For temperatures higher than 200°C heat transfer is only due to forced convection and evaporation of water droplets in the boundary layer. Below 200°C, water droplets were observed as bright spots impacting on the plate surface. They existed on the surface for only a short period of time before evaporating. The

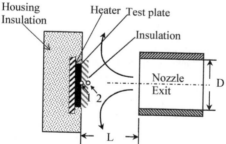

Figure 4: Schematic diagram of furnace housing with 6.4 mm thick insulation over the test plate. Temperatures at points 1 and 2 were measured to determine the heat flux through the insulation on the test plate. Since the heat input is measured, the losses through the housing can be calculated.

number impacting on the plate surface increased as plate temperatures decreased from 200°C to 100°C. Below 100°C, there is no vigorous water boiling and a permanent water film develops on the surface.

In order to obtain the convective heat transfer rate from the measurement of heater power input, the effects of radiation and thermal losses through the heater enclosure must be accounted for. Radiation losses were calculated by assuming the test plate surface to be a grey body with an emissivity that was measured to be between 0.88 and 0.89 over the range of temperatures used. To measure the heater enclosure losses, a set of experiments were conducted with the same procedure as followed in the steady state measurements, except that the test plate was covered with a 6.4 mm thick piece of Kaowool 3000HT ceramic-board insulation as illustrated in Figure 4. For the two-phase flow experiments, a thin copper plate less than 1mm thick was added to protect the insulation from the water.

For a given steady state temperature, the heater enclosure losses can be calculated knowing the heater input power and the heat transfer through the insulated test plate. In order to measure the heat transfer through the insulated test plate, two thermocouples were placed on the outer side of the plate insulation, one at the center and the other on the edge, to obtain an average outside temperature. Since the thermal conductivity of this insulation is known and the temperature at both sides are known, the steady state heat loss through the plate insulation can be calculated from Equation 1 assuming a linear temperature profile.

$$q_{insul} = \frac{k_i \cdot A \cdot (T_{in} - T_{out})}{t} \; ; \quad where \; t = insulation \; thickness \tag{1}$$

Under steady state, the energy input from the heater is either lost through the plate insulation or through the heater enclosure. From the knowledge of the measured heat input and equation 1, the heater enclosure losses are:

$$q_{losses} = q_{input} - q_{insul} \tag{2}$$

Curves of heater enclosure losses can be obtained by varying the heater input, air velocity and water flow rate. Examples of these heater enclosure loss curves are shown in Figure 5. Heater enclosure losses were measured to be between 8 and 43% of the heat input for single-phase flow, with the lower percentages occurring at higher velocities. For two-phase flows, the heater enclosure losses were between 10 and 24% of the heat input, with the lower percentages occurring at high water/air mass flow ratios.

These results suggest that the heater enclosure losses are more important for lower heat transfer rates. For a given plate temperature, as the heat transfer through the plate to the flow increases, the heater enclosure losses increase slowly and become less significant. For single-phase flows, the curves obtained for heater enclosure losses are more linear at lower temperatures and more curved at higher temperatures due to radiation. Heater enclosure losses increase as the air velocity increases at a constant plate steady state temperature, or as the plate temperature increases for a constant air velocity. For two-phase flows the heat losses increase as the water/air mass flow ratio increases for a given steady state temperature.

The steps followed to correct the preliminary steady state heat transfer results are shown in Figure 6 for the case of 3.9m/s. The top curve shows the heater input per unit plate area required to achieve a given steady state temperature. The heater input represents the net power input delivered to the heater by the variac. As expected, the steady state temperature increases as the heater input increases. For other air velocities, the slope of the uncorrected steady state temperature curve decreases with increasing nozzle air velocity.

The measured heater enclosure loss curves (such as the ones shown in Figure 5) were used to calculate a correction for each steady state temperature and flow condition. This correction was used to obtain the middle curve in Figure 6, which is corrected for all losses except radiation. The corrected curve becomes straighter, even though it is not a straight line. The heater enclosure losses were calculated to be between 32 and 33% of the heater input.

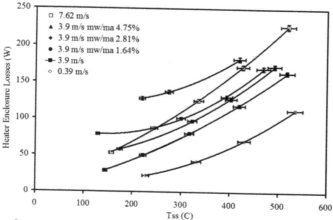

Figure 5: Change in enclosure heat loss with the steady state temperature of the plate. Curve fits of this data were used to correct heater input for enclosure losses.

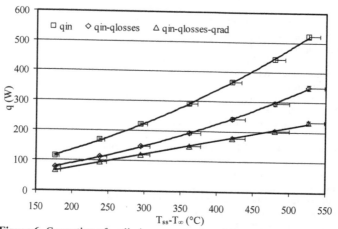

Figure 6: Correction of preliminary results for a 3.9m/s single-phase flow.

The bottom curve is the final result after correcting for radiation loss from the test plate. This curve is a straight line, implying that the convective heat transfer coefficient, which is the slope of this curve, does not change with temperature. Radiation becomes more significant at higher temperatures. It was calculated to be between 8 and 22% of the heater input for this condition, where higher percentages correspond to higher temperatures.

Figure 7 shows the same corrections for a two-phase flow with a water/air mass flow ratio of 4.75% at 3.9m/s. The top curve corresponds to the preliminary results. Like the single-phase flow, as the heater input increases steady state temperature rises, however, the temperature is lower for the two-phase flow. This augmentation of heat transfer is attributed to latent heat of evaporation of water droplets within the boundary layer. As the water/air mass flow ratio increases, the steady state temperature decreases for the same heater input.

The middle curve in Fig. 7 is corrected for all heat losses from the test plate except radiation. For two-phase flows, heater enclosure losses were found for temperatures higher than 200°C, since below that temperature droplets are known to impact the surface and change the

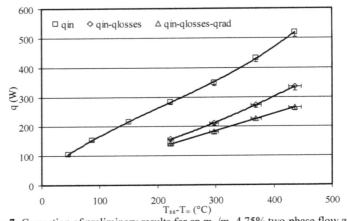

Figure 7: Correction of preliminary results for an m_w/m_a 4.75% two-phase flow at 3.9m/s.

physics of the heat transfer phenomenon from drop evaporating to film cooling. Data taken for steady state temperatures below 200°C for two-phase flows were not used for the calculation of the final results. As with the single-phase flows, the corrected curve becomes straighter. The heater enclosure losses were calculated to be between 35 and 44% of the heater input.

The bottom curve is the final result, after being corrected for radiation from the test plate surface. As expected, the curve is a straight line. Radiation losses were calculated to be between 5 and 13% of the heater input for this condition. Compared to the radiation losses for single-phase flows, they represent a lower percentage due to the lower temperatures reached for two-phase flows.

The top curve in Fig. 7 has an inflection point around 200°C that is attributed to the water droplets impacting on the surface of the test plate. The region around the inflection point is a transition region and ranges from the temperature for which water droplets start to impact on the surface down to the temperature for which a water film develops on the surface. It was observed that when the surface temperature decreases below a certain value water droplets impact on the surface, and heat transfer is enhanced due to the localized quenching effect of water droplets boiling on the surface causing a change in the curvature of the steady state heat flux curve.

Transient Cooling

Transient cooling rates were measured to study the effect of adding droplets. The cooling time is defined as the time to cool the test plate from 500°C to 35°C. These measurements were made with both the single-phase and two-phase flows. A shutter was used to close the nozzle exit to avoid any flow impingement on the test plate. The test plate was heated to 500°C with a constant power input of 310 Watts. The same power input was used for each experiment to ensure that, once the plate reached the desired temperature, the insulation and the heater reached the same conditions for every test. Once the plate was heated up to 500°C, the shutter was opened and the heater was turned off to start the cooling process.

Velocity Field

Several pairs of images were recorded by the PIV system for each case and the time between laser pulses was changed to obtain the correct velocity resolution. Figure 8 shows the velocity vectors for the 3.9m/s and $m_w/m_a = 4.75\%$. The flow is directed upward, the test plate is located above the image and the nozzle exit is located at the bottom of the image. The test plate was not imaged, to avoid the intense laser scattering and enhance the contrast of the light scattered from the particles.

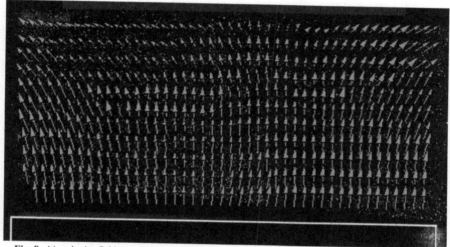

Fig. 8: Air velocity field measured in a single-phase stagnation point flow. Jet exit velocity is 3.9m/s.

PIV measurements were made to determine the optimum distance, L, between plate and nozzle for the experiments. The correct distance ensures that the flow is not affected by the test plate being too close to the nozzle exit or by entrainment of the outside air if the test plate is too far from the nozzle exit. For a measured free exit flow velocity of 3.9m/s, different L/D distances were studied using water droplets as the seeding particles and a water/air mass flow ratio of 2.68%. The results showed that for L/D distances of 0.125, 0.25 and 0.375 the maximum velocity range found at the exit of the nozzle for several pairs of images was less than the free exit velocity of 3.9m/s. However, for an L/D=0.5 the range was $3.8 \leq V_{max} \leq 3.9$m/s and the velocity profile was straight and perpendicular to the nozzle exit. Therefore, it was concluded that the flow at the exit was not affected by the plate for L/D distances greater than 0.5. Measurements for L/D distance ratios greater than 0.5 show that the flow becomes unsteady; therefore L/D of 0.5 was used.

Single-phase flow PIV measurements were made for three exit velocities: 7.62, 3.9 and 0.39m/s. Each resulted in a steady stagnation point flow, and had a uniform velocity profile at the exit, perpendicular to the test plate. The flow field at the exit does not show any large scale turbulent structures that may represent entrainment of outside air. The stagnation point was steady and remained at the center of the hot plate. For the 7.62 m/s exit velocity, the measured velocities ranged from 7.62 m/s at the nozzle exit to 0.1 m/s near the hot plate surface. Similar results were obtained for the addition of water drops.

Particles Size Distribution

Figure 9 shows the drop size distributions resulting from 1380 drops measured in the 3.9 m/s flow with 2.75% water addition. Azzopardi [13] gives the error in the Sauter mean diameter based on this sample size to be approximately ±10%. The droplet flux was 308,000 drops/sec/cm^2 and the distribution consists of a large number of drops in the 20-40μm range and a few large drops in the 70μm and greater range. The most probable diameter is 25μm, the mean diameter is 30μm, and the Sauter mean diameter is 53μm. Half the volume is carried by drops greater than 55μm and 50% of all drops diameters are less than ~22μm.

Bacholo [14] gives the Stoke's law relaxation time for particle Reynolds numbers << 1 as $\tau = d_p^2 \rho_p / 18 v \rho$, where d_p is the particle diameter, ρ_p is the particle density, v is the air kinematic viscosity and ρ is the air density. The Stokes number for this flow can now be defined as $St = L / \tau U_{exit}$, where U_{exit} is the nozzle exit velocity and L is the distance from the exit to the plate. For the 3.9 m/s case, the Stokes numbers are 4.8, 1.0 and 0.3 for drop diameters of 30μm, 65μm and 120μm respectively. Stokes numbers much greater than 1 indicate a droplet that will follow the air flow, therefore for the 3.9 m/s case with 2.75% water addition, drops greater than 65μm are expected to impact on the plate. These drops account for approximately 44% of the volume flow rate and 1.6% of the number of drops. Since droplets only impacted the plate when surface temperatures were less than 200°C (according to visual observations), it is anticipated that for plate temperatures greater than 200°C a portion of these drops is evaporated by heat transfer from the plate in the boundary layer and the resulting smaller drops are swept away with the air flow. The diameter of all the drops is reduced due to evaporation in the boundary layer resulting in an increase in the Stokes number which helps avoid surface impact.

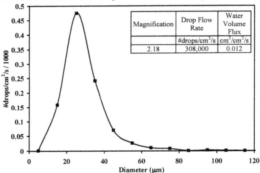

Magnification	Drop Flow Rate	Water Volume Flux
	#drops/cm^2/s	cm^3/cm^2/s
2.18	308,000	0.012

Fig. 9: Water droplet distribution at jet exit for 2.75% water addition, 3.9 m/s. 50% of the water volume is carried by drops with a diameter greater than 55μm. Sauter mean diameter is 52μm and the mean dia. is 30μm. Half of all drops have diameters between ~10 & 22μm and account for 13% of the water flow rate.

RESULTS AND DISCUSSION

Nusselt Number Correlations for Single-Phase Flows

Figure 10 shows the effect of the steady state plate temperature on the plate heat transfer to the flow. The steady state temperature has a linear relationship with the heat flux, which means that the heat transfer coefficient remains constant for the range of temperatures studied. This relationship is characterized by the following equation:

$$q''_{plate} = \frac{q_{plate}}{A} = h \cdot (T_{ss} - T_\infty) \tag{3}$$

Here T_∞ is the ambient temperature, h is the heat transfer coefficient and A is the exposed area of the plate. This equation is used with the corrected steady state plate temperature curves to find the heat transfer coefficient. Once the heat transfer coefficient is found, the Nusselt number Nu = hD/k is calculated, where D is the diameter of the nozzle.

One of the most important relationships for single-phase flow studies is the effect of Reynolds number on the heat transfer rate. The Reynolds number Re = U_{exit}D/ν was calculated for each flow velocity. Figure 11 shows the effect of the Reynolds number on single-phase heat

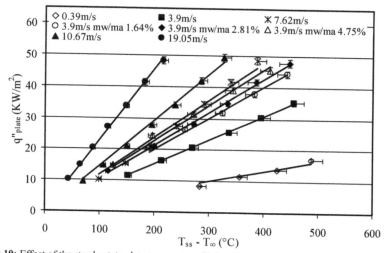

Figure 10: Effect of the steady state plate temperature (T_{ss}) on convective heat flux from the plate ($q"_{plate}$)

transfer. For the lowest velocity used in this study (0.39m/s) the flow is close to the jet turbulent transition zone (Re=2000). As the Reynolds number increases the heat transfer coefficient increases, however, for Re < 20,000 the dependence changes – resulting in weaker dependence for lower velocities and regimes closer to the transition zone. This means that as the Reynolds number decreases within this range the Nusselt number decreases at a lower rate.

The average Nusselt number has been found to have the functional form: $\overline{Nu} = C \cdot Re^m \cdot Pr^n$. The coefficient 'C' depends on the flow characteristics. For jet flows with equivalent exit diameter 'D', 'C' depends on the ratio of nozzle-to-plate distance and nozzle-diameter, L/D. For single round nozzles, Martin [15] and Beitelmal and Saad [16] suggested Nusselt number correlations proportional to $Re^{0.5}$ and $Re^{0.703}$ respectively. Mohanty and Tawfek [17] found Nusselt number dependences between $Re^{0.67}$ and $Re^{0.701}$ for different nozzle diameters. In the current study, for Re > 20,000, the following correlation can be used to represent the data for single-phase flow:

$$\overline{Nu} = 0.2681 \times Re^{0.68}$$

(4)

The power law for the Reynolds number is tabulated for the various studies in Table 1. The power law for the current study agrees well with most of the other studies. Martin's correlation, which does not agree, is only recommended for L/D > 2 and is a result applicable to a wide range of Prandtl and Reynolds numbers. The current correlation does not extend to

Reynolds numbers below 20,000, whereas, Martin's correlation extends down to Re = 2,000. As indicated in Fig.11, if Re = 2,000 was included the value for m would decrease as well.

The Nusselt numbers found earlier by other investigators are lower than those found in the present study. This discrepancy is attributed primarily to the uniform plate temperatures used in this study which resulted from the choice of flow field and heat arrangement. The shorter L/D distance and smaller test plate to jet diameter ratio used in the current study resulted in uniform plate temperatures that more accurately simulate uniform flow approaching from infinity forming a stagnation point. The current flow was arranged to simulate uniform flow approaching infinity, therefore, the L/D was held constant and much smaller than other studies. The Nusselt # of the current study is only slightly larger than the Nusselt # at the stagnation point of Mohanty and Tawfek's [17] study.

Table 1: Single-Phase Flow Studies

Study	Power Law	Ranges of conditions
Current study	0.68	L/D = 0.5 ; 25,000 < Re <122,000
Martin (1977)	~0.5	L/D > 2 ; 2,000 < Re < 400,000
Beitelmal et al (2000)	0.703	1< L/D < 10 ; 9,600 < Re < 38,500
Mohanty et al (1993)	0.696	L/D > 6 ; 6,900 < Re < 24,900

Another important parameter is the characteristic flow field of each study. Unlike for other studies, the flow field used in the present study is well defined in every region. The velocity profile is uniform at the nozzle exit and the flow is perpendicular to the plate surface outside the boundary layer. Moreover, parameters such as the turbulence intensity, vortices and mixing of the jet fluid with the quiescent outer fluid need to be quantified for other studies to have a clear comparison among them. These parameters were not measured for the other studies and there are no PIV measurements available to compare with the flow in the present study.

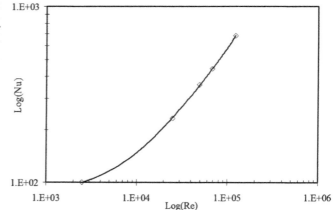

Figure 11: Effect of Reynolds number on single-phase flow heat transfer.

Water Enhancement

Figure 12 shows the effect of water/air mass flow ratio on heat transfer enhancement for a jet exit velocity of 3.9m/s. The heat transfer enhancement is defined as the ratio of the two-phase Nusselt number over the single-phase Nusselt number at the same jet exit velocity. Compressed air is used in the water nozzle to atomize water into small droplets. This additional air flow also has a cooling effect. To quantify the compressed air effect, a set of experiments were performed for single-phase flow adding compressed air from the water nozzle.

Two lines are shown in Fig.12. The upper line corresponds to the two-phase heat transfer enhancement due to the water droplets and the added compressed air. The total heat transfer enhancement effect of the two-phase flow with the compressed air was calculated to be approximately 30, 38 and 48% for water/air mass flow ratios of 1.64, 2.81 and 4.75% respectively. The heat transfer enhancement due only to the water droplets was calculated to be approximately 13, 20 and 31% for water/air mass flow ratios of 1.64, 2.81 and 4.75% respectively.

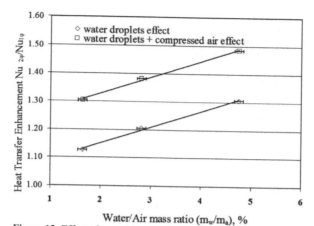

Figure 12: Effect of water/air mass flow ratio on two-phase flow heat transfer enhancement.

Quantification of heat transfer enhancement has been carried out by other investigators. A heat transfer enhancement of 150% was reported for 1.5% mist mass ratio by Li et al [10]. The temperature of the surface reported for Li's experiments was approximately 160°C. The highest heat transfer enhancement reported by Hishida et al [4] was close to 1000%. The surface temperatures used in Hishida's experiments are in the range of 50 to 80°C. The highest local heat transfer enhancement for two-phase flow inside a tube was 300% as reported by Guo et al [18, 19] for 5.1% mist mass ratio and a Reynolds number of 35,000. The surface temperature for Guo's condition was about 200°C. Lee et al [5] reported a maximum of 550% heat transfer enhancement for a channel flow similar to Guo's flow at a velocity of 26.45m/s and 5.05% mist mass ratio. The surface temperature for Lee's case was less than 100°C.

The discrepancy between the heat transfer enhancements obtained from other investigators and those found in the present study is attributed to the different surface temperatures used. For most other studies, the surface temperatures obtained were at or below 200°C, where water droplets reached the surface and created a film. For the water addition in this study, at plate temperatures higher than 200°C the water droplets did not impact on the surface and instead evaporate within the boundary layer.

Transient Cooling

Figure 13 shows the transient cooling behavior for single-phase and two-phase flows. It was observed that a single-phase flow at 3.9m/s takes more than 1800 seconds to cool the plate from 600°C to 35°C. The second curve from the right shows the effect of the compressed air added to the flow. With this excess air, only 1550 seconds are needed to cool the plate to 35°C. For a single-phase flow of 19.05m/s, Fig. 13 indicates that the cooling time is less than 800 seconds, i.e. about twice as fast. As the single-phase flow velocity increases, the heat transfer rate increases, which results in a decrease of the cooling time. The single phase flow transient curves also show a similar exponential decay of the plate temperature where the characteristic time constant decreases with increasing velocity.

For a two-phase flow of m_w/m_a 1.64% the time needed to reach 35°C was 900 seconds. If the water/air mass flow ratio is increased to 2.81% only 600 seconds are needed, and for m_w/m_a 4.75% the cooling time is reduced to 300 seconds, three times lower than that for 1.64%. These results indicate that the water/air mass flow ratio can be used to change the cooling curve.

The transient cooling characteristics of the two-phase flows studied are very different than those of the single-phase flows. As the water/air mass flow ratio increases, the cooling curves change in shape and become straighter. This effect is attributed to changes in heat transfer rate for a two-phase flow in different temperature ranges. The heat transfer rate for temperatures above 200°C is due to convection heat transfer to the flow and evaporation of water droplets within the thermal boundary layer. As the temperature

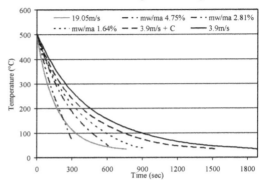

Fig. 13: Single-phase and two-phase flows transient cooling.

decreases, the convection heat transfer rate decreases due to the lower temperature difference between the plate and the environment but the heat transfer rate due to the evaporation of water droplets would remain constant for a given water/mass flow ratio if all the water droplets are evaporated in the boundary layer. However, for temperatures below 200°C the effect of water droplets impinging on the surface enhances the heat transfer rate. For the cases studied, as the temperature of the plate decreases below 200°C more droplets impact on the surface for the same water/air mass flow ratio and the heat transfer rate increases.

For lower water/air mass flow ratios and for temperatures greater than 200°C, the convection heat transfer dominates over droplet impingement heat transfer. Therefore, the cooling curve for lower water/air mass flow ratios is similar to the single-phase flow cooling curves. As the water/air mass flow ratio increases and the plate temperatures decrease, droplet impingement effects become more significant. For m_w/m_a 4.75% the cooling curve is linear, which means that at lower temperatures the droplets impingement effect is high enough to change the overall heat transfer and the result is a constant heat transfer rate throughout the cooling process. These results suggest that the water/air mass flow ratio can be controlled for two-phase flows to meet the requirements of different cooling processes.

CONCLUSIONS
An experimental study was performed to determine the effect of small water droplets incorporated into a single-phase stagnation-point flow on heat transfer. Steady state and transient cooling heat transfer experiments were carried out to analyze the characteristics of single-phase and two-phase flows. PIV measurements were made for different air velocities to determine the characteristics of the flow. Water droplets size distribution was also measured. The following conclusions can be made from this study:
1. The steady state heat transfer measurements and visual observations suggest that above 200°C the two-phase flow heat transfer is due to forced convection and evaporation of water droplets within the thermal boundary layer. Below 200°C, water droplets impinge on the surface and increase the heat transfer rate. As the temperature decreases further, the quantity

of water droplets impinging on the surface increases and the heat transfer rate continues to increase.

2. For the range of water addition studied, the heat transfer enhancement increased linearly with the water/air mass flow ratio. A maximum heat transfer enhancement of 31% was found for a water/air mass flow ratio of 4.75% in the limit of no droplet impingement or surface wetting.

3. The transient cooling characteristics of the two-phase flows studied are very different than those of the single-phase flows. As the water/air mass flow ratio increases, the cooling curves change in shape and become straighter indicating that droplet impingement effects increase the heat transfer rate. The water/air mass flow ratio can be controlled for two-phase flows to modify the characteristic cooling curve and match a specific requirement of a certain industrial cooling process.

ACKNOWLEDGMENTS

The authors wish to thank the Department of Energy's Industrial Assessment Center (Award# DE-EE0005526) and NSF (Award# 1403339) for supporting this work.

REFERENCES

[1] Sozbir, N. and Yao, S.C., 2002, "Experimental Studies of Using Water Mist Cooling for the Tempering of Glass," Proceedings of the ASME International Mechanical Engineering Congress & Exposition, IMECE2002-32524.

[2] Heyt, J. and Larsen, P., 1971, "Heat Transfer to Binary Mist Flow," International Journal of Heat and Mass Transfer, 14(9), pp. 1395-1405.

[3] Toda, S., 1972, "Study of Mist Cooling," Heat Transfer - Jap Res, 1(3), pp. 39-50.

[4] Hishida, K., Maeda, M. and Ikai, S., 1980, "Heat Transfer from a Flat Plate in Two-Component Mist Flow," Journal of Heat Transfer, Transactions ASME, 102(3), pp. 513-518.

[5] Lee, S.L., Yang, A.H. and Hsyua, Y., 1994, "Cooling of a Heated Surface by Mist Flow," Journal of Heat Transfer, Transactions ASME, 116(1), pp. 167-172.

[6] Ohkubo, H. and Nishio S., 1988, "Mist Cooling for Thermal Tempering of Glass," Nippon Kikai Gakkai Ronbunshu, B Hen, 54(501), pp. 1163-1169.

[7] Liu, Z.H. 2002, "Experimental Study of the Boiling Critical Heat Flux of Mist Cooling," Experimental Heat Trasnfer 15:229-243.

[8] Graham, K. and Ramadhyani, S., 1996, "Experimental and Theoretical Studies of Mist Jet Impingement Cooling," Journal of Heat Transfer, Transactions ASME, 118(2), pp. 343-349.

[9] Yao, S.C., Choi, K.J. 1987, "Heat Transfer Experiments of Mono-dispersed Vertically Impacting Sprays." International Journal of Multiphase Flow 13, 639-648.

[10] Li X., Gaddis, J.L. and Wang, T., 2001, "Mist/Steam Heat Transfer in Confined Slot Jet Impingement," Journal of Turbomachinery, 123(1), pp. 161-167.

[11] Everest, D. and Atreya, A., 1999, "Particle Tracking Velocimetry Measurements in Large-Scale Sprinkler Flows," Proceedings of the 2nd Pacific Symposium on Flow Visualization and Image Processing, Honolulu, HI, paper no. PF0126.

[13] Azzopardi, B.J., 1979, "Measurement of Drop Sizes," International Journal of Heat and Mass Transfer, 22, pp.1245-1279.

[14] Bacholo, W.D., 1994, "Experimental Methods in Multiphase Flow," International Journal of Multiphase Flow, 20, Suppl., pp. 261-295.

[15] Martin, H., 1977, "Heat and Mass Transfer Between Impinging Gas Jets and Solid Surfaces," Advances in Heat Transfer, 13.

[16] Beitelmal, A.H. and Saad, M.A., 2000, "Effects of Surface Roughness on the Average Heat Transfer of an Impinging Air Jet," International Communications in Heat and Mass Transfer, 27(1), pp. 1-12.

[17] Mohanty, A.K. and Tawfek A.A., 1993, "Heat Transfer due to a Round Jet Impinging Normal to a Flat Surface," International Journal of Heat and Mass Transfer, **36**(6), pp. 1639-1647.

[18] Guo, T., Wang, T. and Gaddis, J.L., 2000, "Mist/Steam Cooling in a Heated Horizontal Tube – Part I," Journal of Turbomachinery, Transactions of the ASME, 122(2), pp. 360-365.

[19] Guo, T., Wang, T. and Gaddis, J.L., 2000, "Mist/Steam Cooling in a Heated Horizontal Tube – Part II," Journal of Turbomachinery, Transactions of the ASME, 122(2), pp. 366-374.

---

* Corresponding Author. Arvind Atreya, Professor Department of Mechanical Engineering, University of Michigan, Ann Arbor, MI 48109-2125, USA. Email: aatreya@umich.edu, Ph# (734) 647-4790, Fax# 647-3170.

** Water mist is also used for fire suppression as a substitute for halon suppression agents. The challenges here are also to increase the cooling rate as much as possible with the constraint of not wetting the electronic surfaces.

Author Index